Prescriptive Method for Insulating Concrete Forms in Residential Construction

Prepared for

U.S. Department of Housing and Urban Development
Office of Policy Development and Research
Washington, DC

and

Portland Cement Association
Skokie, IL

and

National Association of Home Builders
Washington, DC

Prepared by

NAHB Research Center, Inc.
Upper Marlboro, MD

Contract H-21065CA

May 1998

DISCLAIMER

While the material presented in this document has been prepared in accordance with recognized engineering principles and sound judgment, the document should not be used without first securing competent advice regarding its suitability for any given application. **The use of this document is subject to approval by the local code authority.** Although the information in the document is believed to be accurate, neither the authors, nor reviewers, nor the U.S. Department of Housing and Urban Development of the U.S. Government, nor the Portland Cement Association, nor the National Association of Home Builders, nor the NAHB Research Center, nor any of their employees or representatives makes any warranty, guarantee, or representation, expressed or implied, with respect to the accuracy, effectiveness, or usefulness of any information, method, or material in this document, nor assumes any liability for the use of any information, methods, or materials disclosed herein, or for damages arising from such use.

NOTICE

The contents of this report are the views of the contractor and do not necessarily reflect the views or policies of the U.S. Department of Housing and Urban Development or the U.S. government.

FOREWORD

For centuries, home builders in the United States have made wood their material of choice because of its satisfactory performance, abundant supply, and relatively low cost. However, unpredictable fluctuations in price and problems cited with the quality of framing lumber are causing builders to seek innovative building products.

Insulating Concrete Forms (ICFs) represent a category of potential building product that is receiving greater attention among builders. ICFs are hollow blocks, planks, or panels that can be constructed of rigid foam plastic insulation, a composite of cement and foam insulation, a composite of cement and wood chips, or other suitable insulation material that has the ability to act as forms for cast-in-place concrete walls. The forms typically remain in place after the concrete has cured, providing well-insulated construction. ICFs are gaining popularity because they are competitive with light-frame construction and offer a strong, durable, and energy-efficient wall system for housing.

The lack of a consistent and comprehensive set of prescriptive requirements has prevented ICF systems from reaching their full potential among home builders and code officials who may be unfamiliar with this construction technique. As a result, those who desire to build or purchase ICF homes must incur the additional cost burden of engineering for each application. This document represents the outcome of an initial effort to fulfill the need for prescriptive requirements and to improve the overall affordability of homes constructed with insulating concrete forms.

This publication was developed under sponsorship of the U.S. Department of Housing and Urban Development (HUD) through a cooperative agreement with the National Association of Home Builders (NAHB) and the Portland Cement Association (PCA). This publication was written by the NAHB Research Center with assistance from a steering committee, which represents the interests and expertise of ICF manufacturers, ICF producers, code officials, researchers, professional engineers, and builders experienced in ICF construction.

Paul A. Leonard
Deputy Assistant Secretary for Policy Development
U.S. Department of Housing and Urban Development

ACKNOWLEDGMENTS

This report was prepared by the NAHB Research Center under sponsorship of the U.S. Department of Housing and Urban Development (HUD). We wish to recognize the National Association of Home Builders (NAHB) and the Portland Cement Association (PCA) both of whose co-funding and participation made the project possible. Special appreciation is extended to William Freeborne of HUD, Donn Thompson of PCA, and Lionel Lemay, P.E., of PCA for guidance throughout the project.

The principal authors of this report are Andrea Vrankar, R.A., and Nader Elhajj, P.E., with support from Shawn McKee and Stephen Scully, P.E. Technical review was provided by Jay Crandell, P.E. Clerical support was provided by Lynda Marchman, and figures were produced by Barbara Vrankar Karim. Appreciation is especially extended to members of the steering committee (listed below) who provided guidance on this document and whose input made this work complete.

Ron Ardres, Reddi-Form, Inc.
Barney Barnett, Superior Built
Lance Berrenberg, American Polysteel Forms
Pat Boeshart, Lite-Form, Inc.
Jonathan Childres, North State Polysteel
Jay Crandell, P.E., NAHB Research Center, Inc.
Bill Crenshaw, Perma-Form Components, Inc.
Ken Demblewski, Sr, P.E., K and B Associates, Inc.
Nader Elhajj, P.E., NAHB Research Center, Inc.
Anne Ellis, P.E., National Reddi-Mix Concrete Association
William Freeborne, U.S. Department of Housing and Urban Development
Thomas Greeley, BASF Corporation
David Hammerman, P.E., Howard County (Maryland) Department of Inspections, Licenses, and Permits
Bob Hartling, Poly-Forms, LLC
Gary Holland, Perma-Form Components, Inc.
Byron Hulls, Owens-Corning
Raj Jalla, Consulting Engineers Corp.
Lionel Lemay, P.E., Portland Cement Association
Paul Lynch, Fairfax County (Virginia) Department of Inspection Services
Roger McKnight, Romak & Associates, Inc.
Andrew Perlman, Alexis Homes
T. Reid Pocock, Jr., Dominion Building Group, Inc.
Frank Ruff, TopCraft Homes, Inc.
Robert Sculthorpe, AAB Building System, Inc.
Dean Seibert, Avalon Concepts Corp.
Jim Shannon, Huntsman Chemical Corp.
Steven Skalko, P.E., Portland Cement Association
Herbert Slone, Owens-Corning
Glen Stoltzfus, VA Polysteel Wall Systems
Donn Thompson, Portland Cement Association
Stan Traczuk, Avalon Concepts Corp.
Ned Trautman, Owens-Corning
Andrea Vrankar, R.A., NAHB Research Center, Inc.
Hansruedi Walter, K-X Industries, Inc.
Dick Whitaker, Insulating Concrete Form Association
Lee Yost, Advanced Building Structure
Roy Yost, Advanced Building Structure

The NAHB Research Center appreciates and recognizes the following companies that provided ICFs, tools, and other materials to support various research and testing efforts.

American Polysteel Forms
Avalon Concepts, Corp.
Lite-Form, Inc.
Reddi-Form, Inc.

TABLE OF CONTENTS

LIST OF TABLES

PART I: PRESCRIPTIVE METHOD

PART II: COMMENTARY

APPENDIX A: ILLUSTRATIVE EXAMPLE

LIST OF FIGURES

PART I: PRESCRIPTIVE METHOD

APPENDIX A: ILLUSTRATIVE EXAMPLE

APPENDIX B: ENGINEERING TECHNICAL SUBSTANTIATION

EXECUTIVE SUMMARY

The *Prescriptive Method for Insulating Concrete Forms in Residential Construction* was developed as a guideline for the construction of one- and two-family residential dwellings that use insulating concrete form (ICF) systems. It provides a prescriptive method for the design, construction, and inspection of homes that take advantage of ICF technology. This document standardizes the minimum requirements for basic ICF systems and provides an identification system for the different types of ICFs. It specifically includes minimum wall thickness tables, reinforcement tables, lintel span tables, and connection requirements. The requirements are supplemented with appropriate construction details in an easy-to-read format.

The *Prescriptive Method for Insulating Concrete Forms in Residential Construction (Prescriptive Method)* is consistent with the intent of current U.S. building code provisions, engineering standards, and industry specifications, but it is not written as a regulatory document. The *Prescriptive Method* is written in a building code-compatible style to expedite future code adoption.

This document is divided into two parts:

I. Prescriptive Method

The *Prescriptive Method* is a guideline to facilitate the use of ICF wall systems in the construction of one- and two-family residential dwellings. By providing a prescriptive method for the construction of typical homes with ICF systems, the guideline eliminates the need for engineering in most applications. The provisions in this document were developed by applying accepted engineering practices and practical construction techniques; however, users of the document should verify its compliance with local code requirements.

II. Commentary

The *Commentary* is provided to expedite the use of and to provide the necessary background, supplemental information, and engineering assumptions and methods for the *Prescriptive Method.* The individual sections, figures, and tables are presented in the same sequence as in the *Prescriptive Method.*

Three appendices are also provided.

A. Illustrative Example

The *Illustrative Example* contains design examples illustrating the proper application of the different standards and specifications in the *Prescriptive Method.* It provides a step-by-step procedure on how to apply the requirements of the *Prescriptive Method* when designing a typical home.

B. Engineering Technical Substantiation

The *Engineering Technical Substantiation* contains the engineering calculations and assumptions used to generate the wall, lintel, and connection tables in the *Prescriptive Method.*

C. Metric Conversion Factors

The *Metric Conversion Factors* provides the conversion relationship between U.S. customary units and the International System (SI) units. A complete guide to the SI system and its use can be found in ASTM E 380 [1].

PART I

PRESCRIPTIVE METHOD

INTRODUCTION

The *Prescriptive Method* is a guideline to facilitate the use of ICF wall systems in the construction of one- and two-family residential dwellings. By providing a prescriptive method for the construction of typical homes with ICF systems, the guideline eliminates the need for engineering in most applications. While the provisions in this document were developed by applying accepted engineering practices and practical construction techniques, users of this document should verify compliance of the provisions with local code requirements.

This is not a regulatory document, although it is written for that purpose. The user should refer to applicable building code requirements when exceeding the limitations of this document, when requirements conflict with the building code, or when an engineered design is specified. *This specification is not intended to limit the appropriate use of concrete or construction not specifically prescribed. This document is also not intended to restrict the use of sound judgment or exact engineering analysis of specific applications that may result in designs with improved performance and economy.*

Information is presented in both U.S. customary units and International System (SI) units except for reinforcement bar sizes which are only presented in U.S. customary units. Refer to Appendix C for the corresponding reinforcement bar size in SI units.

1.0 GENERAL

1.1 Purpose

This document provides prescriptive requirements for the use of insulating concrete form systems in the construction of residential structures. Included are definitions, limitations of applicability, below-grade and above-grade wall design tables, lintel tables, various construction and thermal guidelines, and other related information for home builders, building code officials, and design professionals.

1.2 Approach

The prescriptive requirements are based primarily on the *Building Code Requirements for Structural Concrete* [2] and the *Structural Design of Insulating Concrete Form Walls in Residential Construction* [3] for member strength and reinforcement requirements. To a lesser extent, the requirements are also based on *Minimum Design Loads for Buildings and Other Structures* [4], the *Standard Building Code* [5], the *Uniform Building Code* [6], the *National Building Code* [7], and the *One- and Two-Family Dwelling Code* [8]. In addition, the requirements incorporate construction practices from the *Guide to Residential Cast-in-Place Concrete Construction* [9].

The provisions represent sound engineering and construction practice, taking into account the need for practical and affordable construction techniques for lightly loaded residential buildings up to two-stories above grade. This document is not intended to restrict the use of sound judgment or exact engineering analysis of specific applications that may result in designs with improved performance and economy. The engineering calculations that form the basis for this document are discussed in Appendix B, *Engineering Technical Substantiation.*

1.3 Scope

The provisions of the *Prescriptive Method* apply to the construction of detached one- and two-family homes, townhouses, and other attached single-family dwellings in accordance with the general limitations of Table 1.1. The limitations are intended to define the appropriate use of this document for most one- and two-family dwellings. An engineered design shall be required for houses built along the immediate, hurricane-prone coastline subjected to storm surge (i.e., beach front property). Intermixing of the present provisions with other construction materials in a single structure shall be in accordance with the applicable building code requirements for that material, the general limitations set forth in Table 1.1, and relevant provisions of this document.

1.4 ICF System Limitations

There are three categories of ICF systems based on the resulting shape of the formed concrete wall. The shape of the concrete wall may be better understood by visualizing the form stripped away from the concrete, thereby exposing it to view as shown in Figure 1.1. The three categories of ICF wall types are (1) flat, (2) grid, and (3) post-and-beam. The grid wall type is further categorized into (2a) waffle-grid and (2b) screen-grid wall systems.

The provisions of this document shall be used for concrete walls constructed with flat, waffle-grid, or screen-grid ICF systems as shown in Figure 1.1, defined in Section 1.5, and in accordance with the limitations of Section 2.0. Other systems, such as post-and-beam, shall be permitted with an approved design and in accordance with the manufacturer's recommendations.

TABLE 1.1
APPLICABILITY LIMITS

ATTRIBUTE	MAXIMUM LIMITATION
GENERAL	
Maximum Building Plan Dimension	60 feet (18 m)
Number of Stories	2 stories above grade plus a basement
Story Height	10 feet (3 m)
Design Wind Speed	110 mph (177 km/hr) fastest-mile wind speed
Ground Snow Load	70 psf (3.4 kPa)
Seismic Zone	0, 1, and 2
FOUNDATIONS	
Unbalanced Backfill Height	9 feet (2.7 m)
Equivalent Fluid Density of Soil	60 pcf (960 kg/m^3)
Presumptive Soil Bearing Value	2,000 psf (96 kPa)
WALLS	
Unit Weight of Concrete	150 pcf (23.6 kN/m^3)
Load-Bearing Wall Height	10 feet (3 m)
FLOORS	
Floor Dead Load	15 psf (0.72 kPa)
First-Floor Live Load	40 psf (1.9 kPa)
Second-Floor Live Load (sleeping rooms)	30 psf (1.4 kPa)
Floor Clear Span (unsupported)	32 feet (9.8 m)
ROOFS	
Roof Slope	12:12
Roof and Ceiling Dead Load	15 psf (0.72 kPa)
Roof Live Load (ground snow load)	70 psf (3.4 kPa)
Attic Live Load	20 psf (0.96 kPa)
Roof Clear Span (unsupported)	40 feet (12 m)

For SI: 1 foot = 0.3048 m; 1 psf = 47.8804 Pa; 1 pcf = 157.0877 N/m^3 = 16.0179 kg/m^3; 1 mph = 1.6093 km/hr

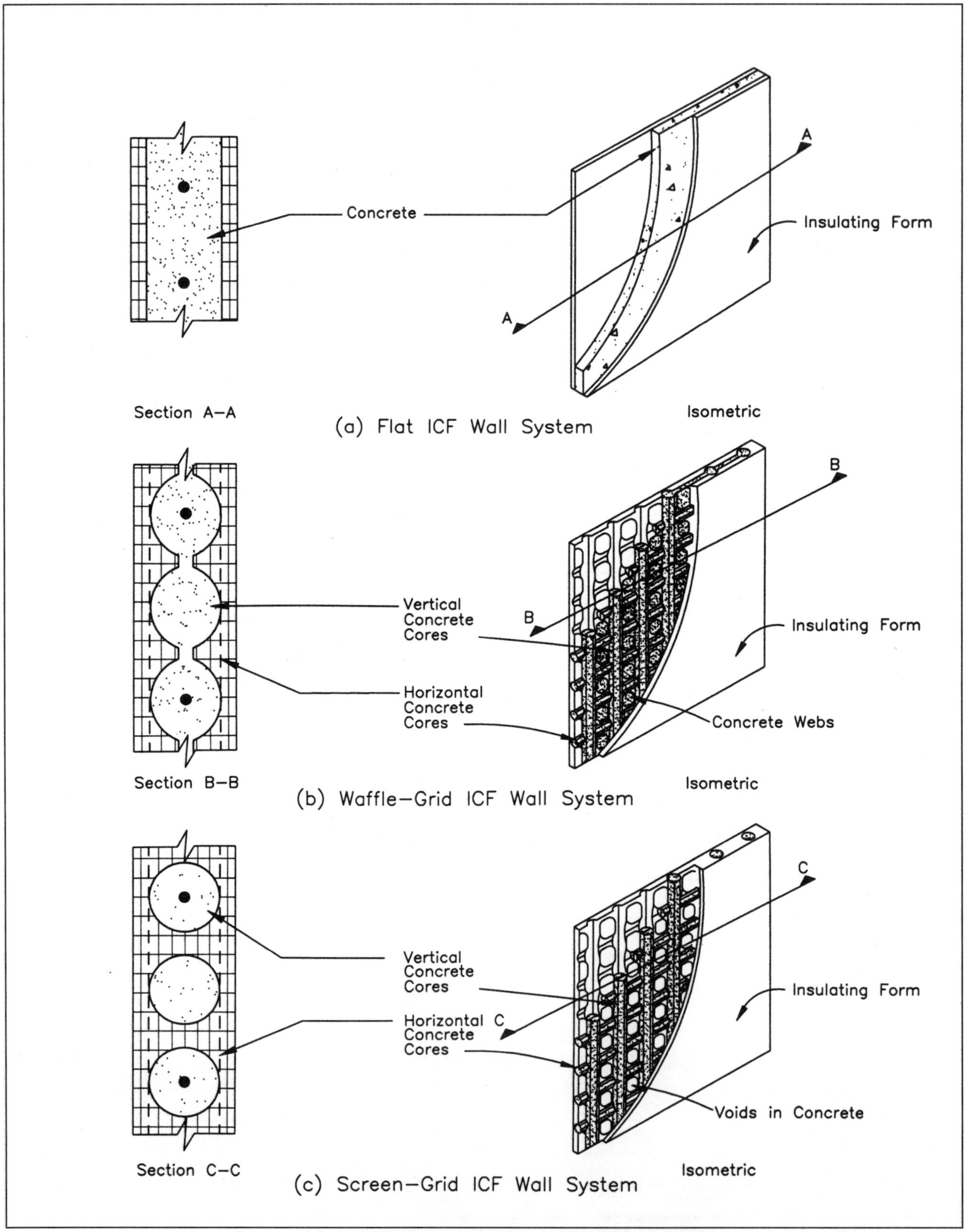

Figure 1.1 ICF Wall Systems Covered by This Document

1.5 Definitions

Accepted Engineering Practice: An engineering approach that conforms with accepted principles, tests, technical standards, and sound judgment.

Anchor Bolt: A bolt, headed or threaded, used to connect a structural member of different material to a concrete member.

Approved: Reference to approval by the building code authority having jurisdiction. A rational design by a competent design professional shall constitute grounds for approval.

Attic: The enclosed space between the ceiling joists of the top-most floor and the roof rafters of a building, not intended for occupancy but sometimes used for storage.

Authority Having Jurisdiction: The organization, political subdivision, office, or individual charged with the responsibility of administering and enforcing the provisions of applicable building codes.

Backfill: The soil that is placed adjacent to completed portions of a below-grade structure (i.e., basement) with suitable compaction and allowance for settlement.

Basement: That portion of a building which is partly or completely below grade and which may be used as habitable space.

Bond Beam: A continuous horizontal beam of concrete with steel reinforcement located in the exterior walls of a structure to tie the structure together and distribute loads.

Buck: A frame constructed of wood, plastic, vinyl, or other suitable material set in a concrete wall opening that provides a suitable surface for fastening a window or door frame.

Building: Any one- or two-family dwelling or portion thereof that is used for human habitation.

Building Aspect Ratio: A building's length divided by its width; used to determine the amount of solid concrete wall required to adequately resist lateral loads from wind and earthquakes forces; refer to Section 5.0.

Building Length: The dimension of a building that is perpendicular to roof rafters, roof trusses, or floor joists. The longer plan dimension, L, of a building when determining the building aspect ratio.

Building Width: The dimension of a building that is parallel to roof rafters, roof trusses, or floor joists. The shorter plan dimension, W, of a building when determining the building aspect ratio.

Cold Joint: A joint or discontinuity resulting from concrete cast against concrete that has already set or cured.

Compressive Strength: The maximum ability of concrete to resist a compressive load, usually measured in pounds per square inch (psi) or Pascals (Pa). The compressive strength is based on compression tests of concrete cylinders that are moist-cured for 28 days in accordance with ASTM C 31 [10] and ASTM C 39 [11].

Crawlspace: A type of building foundation that uses a perimeter foundation wall to create an under floor space which is not habitable.

Dead Load: Forces resulting from the weight of walls, partitions, framing, floors, ceilings, roofs, and all other permanent construction entering into, and becoming part of, a building.

Deflection: Elastic movement of a loaded structural member or assembly (i.e., beam or wall).

Design Professional: An architect or engineer, registered or licensed to practice professional architecture or engineering, as defined by the statutory requirements of the laws of the state in which the project is to be constructed.

Design (or Basic) Wind Speed: Related to winds that are expected to be exceeded once every 50 years at a given site (i.e., 50-year return period). Wind speeds in this document are given in units of miles per hour (mph) by "fastest-mile" measurements. A “standard” wind exposure is assumed in this document in accordance with the SBC [5] to provide design wind loads. The wind loads are adequate for buildings situated in open terrain and on the hurricane-prone coastline; however, they are presumed to be conservative for suburban, urban, and wooded terrain. Wind speed-up from topographic effects (i.e., protruding hills and ridges) are not considered in this document.

Dwelling: Any building that contains one or two dwelling units for living purposes.

Eccentric Load: A force imposed on a structural member at some point other than its center-line, such as the forces transmitted from the floor joists to an exterior wall through a ledger board connection.

Equivalent Fluid Density: The mass of a soil per unit volume treated as a fluid mass for the purpose of determining lateral design loads produced by the soil on an adjacent structure. Refer to the *Commentary* for suggestions on relating equivalent fluid density to soil type.

Flame-Spread Rating: The combustibility of a material that contributes to fire impact through flame spread over its surface; refer to ASTM E 84 [12].

Flat Wall: A solid concrete wall of uniform thickness produced by ICFs or other forming systems Refer to Figure 1.1.

Floor Joist: A horizontal structural framing member that supports floor loads.

Footing: A below-grade foundation component that transmits loads directly and safely to the underlying earth.

Form Tie: The element of an ICF system that holds both sides of the form together. Form ties can be steel, solid plastic, foam plastic, a composite of cement and wood chips, a composite of cement and foam plastic, or other suitable material capable of resisting the loads created by wet concrete. Form ties remain permanently embedded in the concrete wall.

Foundation: The structural elements through which the load of a structure is transmitted to the earth.

Foundation Wall: The structural element of a foundation that transmits the load of a structure to the earth; includes basement, stem, and crawlspace walls.

Grade: The finished ground level adjoining the building at all exterior walls.

Grade Plane: A reference plane representing the average of the finished ground level adjoining the building at all exterior walls.

Ground Snow Load: Measured load on the ground due to snow accumulation developed from a statistical analysis of weather records expected to be exceeded once every 50 years at a given site (i.e., 50-year return period).

Horizontal Reinforcement: Steel reinforcement placed horizontally in concrete walls to provide resistance to temperature and shrinkage cracking. In certain circumstances, horizontal reinforcement is required for additional strength around openings and in high loading conditions such as experienced in hurricanes and earthquakes.

Insulating Concrete Forms (ICFs): A concrete forming system using stay-in-place forms of foam plastic insulation, a composite of cement and foam insulation, a composite of cement and wood chips, or other insulating material for constructing cast-in-place concrete walls. Some systems are designed to have one or both faces of the form removed after construction.

Interpolation: A mathematical process used to compute an intermediate value of a quantity between two given values assuming a linear relationship.

J Bolt: A threaded anchor bolt typically embedded in concrete with threads on one end and a crook in the shank at the other; used to connect a structural member of different material to a concrete member.

Lap Splice: Formed by extending reinforcement bars past each other a specified distance to permit the force in one bar to be transferred by bond stress through the concrete and into the second bar. Permitted when the length of one continuous reinforcement bar is not practical for placement.

Lateral Load: A horizontal force, created by wind or earthquake, acting on a structure or its components.

Lateral Support: A horizontal member providing stability to a column or wall across its smallest dimension. Walls designed in accordance with Section 5.0 provide lateral stability to the whole building when experiencing wind or earthquake events.

Ledger: A horizontal structural member fastened to a wall to serve as a connection point for other structural members, typically floor joists.

Lintel: A horizontal structural element of reinforced concrete located above an opening in a wall to support the construction above.

Live Load: Any gravity load that is not permanently applied to a structure; typically transient and sustained gravity forces resulting from the weight of people and furnishings, respectively.

Load-Bearing Value of Soil: The allowable load per surface area of soil without failing. It is usually expressed in pounds per square foot (psf) or Pascals (Pa).

Post-and-Beam Wall: A perforated concrete wall with widely spaced (greater than that required for screen-grid walls) vertical and horizontal concrete members (cores) with voids in the concrete between the cores created by the ICF form. The post-and-beam wall resembles a concrete frame rather than a monolithic concrete (i.e., flat, waffle-, or screen-grid) wall and requires a different engineering analysis per ACI 318 [2]; therefore, it is not addressed in this edition of the *Prescriptive Method.*

Presumptive: Formation of a judgment on probable grounds until further evidence is received.

R-Value: Coefficient of thermal resistance. A standard measure of the resistance that a material offers to the flow of heat; it is expressed as $\frac{{}^{o}F \cdot hr \cdot ft^2}{Btu}$.

Roof Snow Load: Uniform live load on the roof due to snow accumulation; roughly equivalent to 70 to 80 percent of the ground snow load in accordance with ASCE 7 [4].

Screen-Grid Wall: A perforated concrete wall with closely spaced vertical and horizontal concrete members (cores) with voids in the concrete between the members created by the ICF form; refer to Figure 1.1. It is also called an interrupted-grid wall or post-and-beam wall in other publications.

Seismic Load: The force exerted on a building structure resulting from seismic (earthquake) ground motions.

Seismic Zones: Designated areas associated with a particular level or range of seismic risk and associated seismic design parameters (i.e., peak ground acceleration). Seismic Zones 0, 1, and 2 correspond to successively greater seismic design loads. The higher Seismic Zones 3 and 4 are excluded from this edition of the *Prescriptive Method.*

Sill Plate: A horizontal member constructed of wood, vinyl, plastic, or other suitable material that is fastened to the top of a concrete wall, providing a suitable surface for fastening structural members constructed of different materials to the concrete wall.

Slab-on-Grade: A concrete floor which is supported by, or rests on, the soil directly below.

Slump: A measure of consistency of freshly mixed concrete equal to the amount that a cone of uncured concrete sags below the mold height after the cone-shaped mold is removed in accordance with ASTM C 143 [13].

Smoke-Development Rating: The combustibility of a material that contributes to fire impact through life hazard and property damage by producing smoke and toxic gases; refer to ASTM E 84 [12].

Span: The clear horizontal or vertical distance between supports.

Stem Wall: A below-grade foundation wall of uniform thickness supported directly by the soil or on a footing. Wall thickness and height are determined as that which can adequately distribute the building loads safely to the earth.

Stirrup: Steel bars, wires, or welded wire fabric located perpendicular to horizontal reinforcement and extending across the depth of the member in concrete beams, lintels, or similar members subject to large shear loads.

Story: That portion of the building included between the upper surface of any floor and the upper surface of the floor next above, except that the top-most story shall be that habitable portion of a building included between the upper surface of the top-most floor and the ceiling or roof above.

Story Above-Grade: Any story with its finished floor surface entirely above grade except that a basement shall be considered as a story above-grade when the finished surface of the floor above the basement is (a) more than 6 feet (1.8 m) above the grade plane, (b) more than 6 feet (1.8 m) above the finished ground level for more than 50 percent of the total building perimeter, or (c) more than 12 feet (3.7 m) above the finished ground level at any point.

Structural Fill: An approved, noncohesive material such as crushed rock or gravel.

Townhouse: Single-family dwelling unit constructed in a row of attached units separated by fire walls at property lines and with open space on at least two sides.

Unbalanced Backfill Height: Typically the difference between the interior and exterior finish grade. Where an interior concrete slab is provided, the unbalanced backfill height is the difference in height between the exterior finish grade and the interior floor or slab surface of a basement or crawlspace.

Vapor Retarder: A layer of material used to retard the transmission of water vapor through a building wall or floor.

Vertical Reinforcement: Steel reinforcement placed vertically in concrete walls to strengthen the wall when heavily loaded, especially when lateral forces or large eccentric loads are present. In certain circumstances, vertical reinforcement is required for additional strength around openings.

***Waffle-Grid Wall*:** A solid concrete wall with closely spaced vertical and horizontal concrete members (cores) with a concrete web between the members created by the ICF form; refer to Figure

1.1. The thicker vertical and horizontal concrete cores and the thinner concrete webs create the appearance of a breakfast waffle. It is also called an uninterrupted-grid wall in other publications.

Wall Height: The clear vertical distance between the finished floor and the finished ceiling. Where a finished floor does not exist (i.e., crawlspace), the wall height is the clear vertical distance between the interior finish grade and the finished ceiling.

Web: A concrete wall segment, a minimum of 2 inches (51 mm) thick, connecting the vertical and horizontal concrete members (cores) of a waffle-grid ICF wall or lintel member. Webs may contain form ties but are not reinforced (i.e., vertical or horizontal reinforcement or stirrups). Refer to Figure 1.1.

Wind Load: The force or pressure exerted on a building structure and its components resulting from wind. Wind loads are typically measured in pounds per square foot (psf) or Pascals (Pa).

Yield Strength: The ability of steel to withstand a tensile load, usually measured in pounds per square inch (psi) or Pascals (Pa). It is the highest tensile load that a material can resist before permanent deformation occurs as measured by a tensile test in accordance with ASTM A 370 [14].

2.0 MATERIALS, SHAPES, AND STANDARD SIZES

2.1 Physical Dimensions

Concrete walls constructed with ICF systems in accordance with this document shall comply with the shapes and minimum concrete cross-sectional dimensions required in this section. ICF systems resulting in concrete walls not in compliance with this section shall be used in accordance with the manufacturer's recommendations and as approved.

2.1.1 Flat ICF Wall Systems

Flat ICF wall systems shall comply with Figure 2.1 and shall have a minimum concrete thickness of 5.5 inches (140 mm) for basement walls and 3.5 inches (89 mm) for above-grade walls.

2.1.2 Waffle-Grid ICF Wall Systems

Waffle-grid ICF wall systems shall have a minimum nominal concrete thickness of 6 inches (152 mm) for the horizontal and vertical concrete members (cores). The actual dimension of the cores shall comply with the dimensional requirements of Table 2.1 and Figure 2.2.

2.1.3 Screen-Grid ICF Wall System

Screen-grid ICF wall systems shall have a minimum nominal concrete thickness of 6 inches (152 mm) for the horizontal and vertical concrete members (cores). The actual dimensions of the cores shall comply with the dimensional requirements of Table 2.2 and Figure 2.3.

2.2 Concrete Materials

2.2.1 Concrete Mix

Ready-mixed concrete for ICF walls shall meet the requirements of ASTM C 94 [15]. Maximum slump shall not be greater than 6 inches (152 mm) as determined in accordance with ASTM C 143 [13]. Maximum aggregate size shall not be larger than 3/4 inch (19 mm).

> **Exception:** Maximum slump requirements may be exceeded for approved concrete mixtures resistant to segregation, meeting the concrete compressive strength requirements, and in accordance with the ICF manufacturer's recommendations.

2.2.2 Compressive Strength

The minimum compressive strength of concrete, f_c', shall be 2,500 psi (17.2 MPa) at 28 days as determined in accordance with ASTM C 31 [10] and ASTM C 39 [11].

2.2.3 Reinforcing Steel

Reinforcing steel used in ICFs shall meet the requirements of ASTM A 615 [16], ASTM A 616 [17], ASTM A 617 [18], or ASTM A 706 [19]. The minimum yield strength of the reinforcing steel shall be Grade 40 (300 MPa). Reinforcement shall be secured in the proper location in the forms with tie wire or other bar support system such that displacement will not occur during the concrete placement operation. Steel reinforcement shall have a minimum 3/4-inch (19-mm) concrete cover. Horizontal and vertical wall reinforcement shall not vary outside of the middle third of beams, columns, lintels, horizontal and vertical cores, and flat walls for all wall sizes.

Exception: Horizontal and vertical wall reinforcement in basement and crawlspace walls is permitted to be placed closer to the inside face of the wall provided that it does not conflict with the minimum required concrete cover of 3/4 inch (19 mm).

Vertical and horizontal wall reinforcement required in Sections 3.0, 4.0, and 5.0 shall be constructed from the longest lengths practical. Where joints occur in vertical and horizontal wall reinforcement, a lap splice shall be provided in accordance with Figure 2.4.

Except where otherwise noted, lap splices shall be a minimum of $40d_b$ in length, where d_b is the diameter of the smaller bar. The maximum gap between noncontact parallel bars at a lap splice shall not exceed $d_b/8$, where d_b is the diameter of the smaller bar.

2.3 Form Materials

Insulating concrete forms shall be constructed of rigid foam plastic meeting the requirements of ASTM C 578 [20], a composite of cement and foam insulation, a composite of cement and wood chips, or other approved material. Forms shall provide sufficient strength to contain concrete during the concrete placement operation. Flame-spread rating of forms shall be less than 75 and smoke-developed rating of forms shall be less than 450 tested in accordance with ASTM E 84 [12].

TABLE 2.1
DIMENSIONAL REQUIREMENTS FOR CORES AND WEBS IN WAFFLE-GRID ICF WALLS[1]

Nominal Size inches (mm)	Minimum Width of Vertical Core, W inches (mm)	Minimum Thickness of Vertical Core, T inches (mm)	Maximum Spacing of Vertical Cores inches (mm)	Maximum Spacing of Horizontal Cores inches (mm)	Minimum Web Thickness inches (mm)
6 (152)	6.25 (159)	5 (127)	12 (305)	16 (406)	2 (51)
8 (203)	7 (178)	7 (178)	12 (305)	16 (406)	2 (51)

[1] Width "W", thickness "T", and spacing are as shown in Figure 2.2.

TABLE 2.2
DIMENSIONAL REQUIREMENTS FOR CORES IN SCREEN-GRID ICF WALLS[1]

Nominal Size inches (mm)	Minimum Width of Vertical Core, W inches (mm)	Minimum Thickness of Vertical Core, T inches (mm)	Maximum Spacing of Vertical Cores inches (mm)	Maximum Spacing of Horizontal Cores inches (mm)
6 (152)	5.5 (140)	5.5 (140)	12 (305)	12 (305)

[1] Width "W", thickness "T", and spacing are as shown in Figure 2.3.

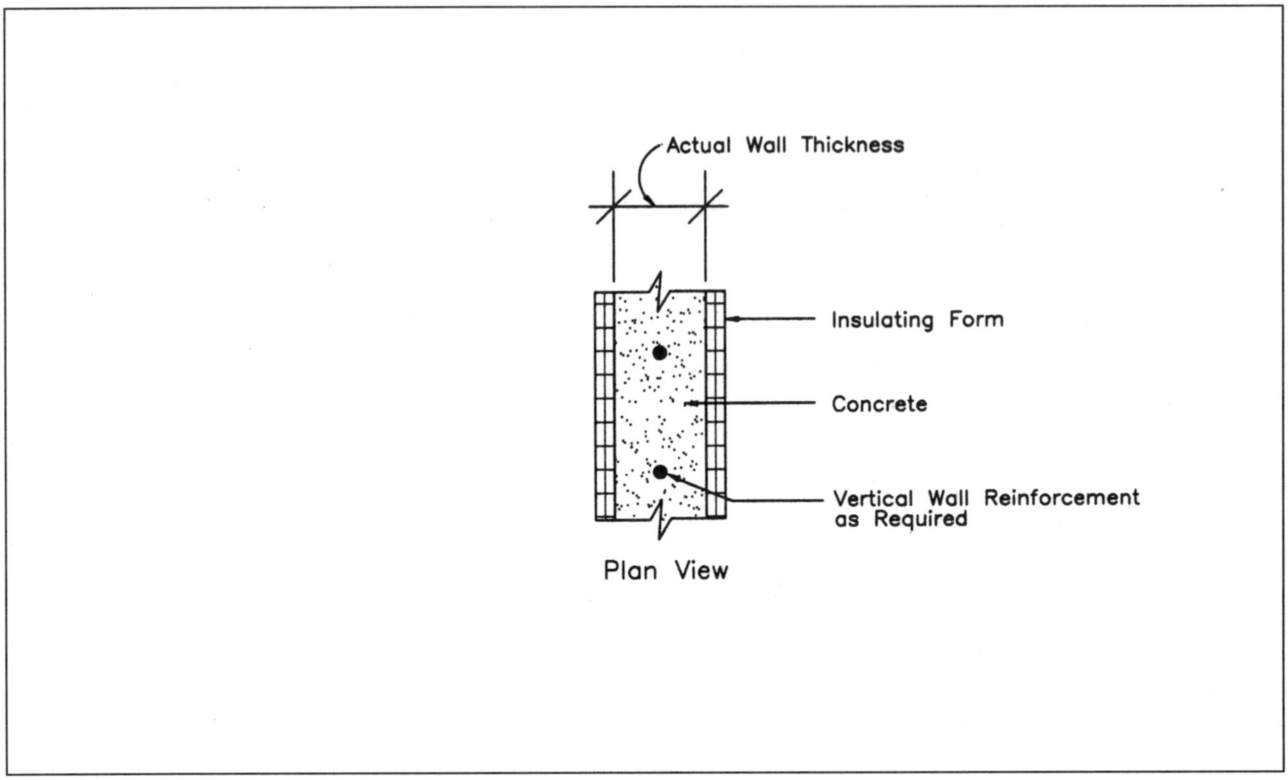

Figure 2.1 Flat ICF Wall System Requirements

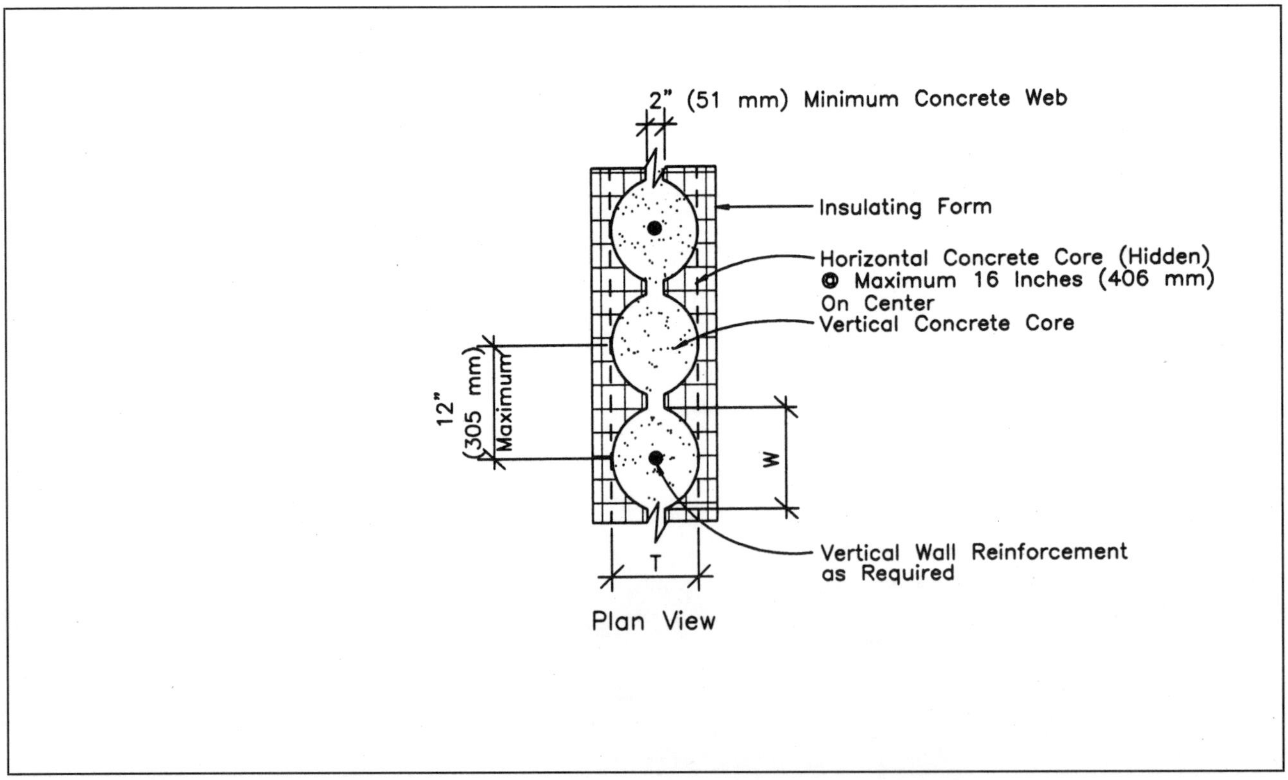

Figure 2.2 Waffle-Grid ICF Wall System Requirements

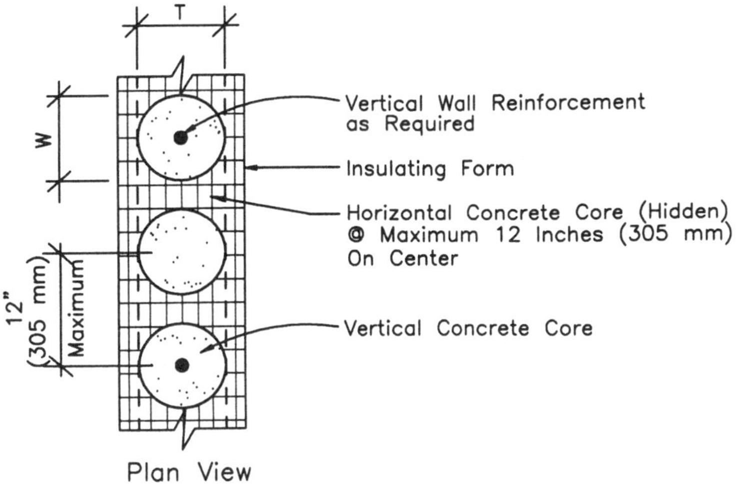

Figure 2.3 Screen-Grid ICF Wall System Requirements

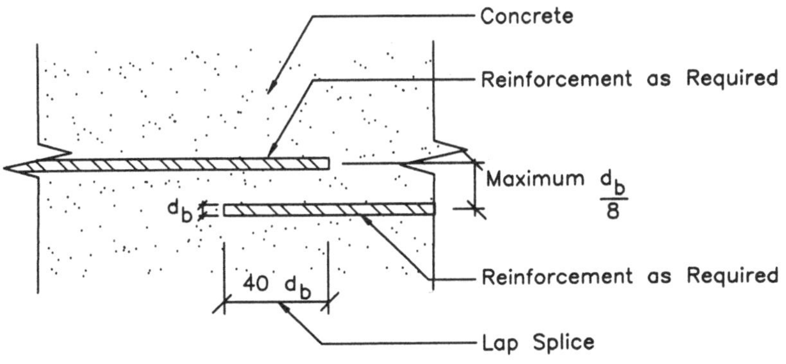

Figure 2.4 Lap Splice Requirements

3.0 FOUNDATIONS

3.1 Footings

All exterior ICF walls shall be supported on continuous ICF, solid masonry, or concrete footings, or other approved systems of sufficient design to safely transmit the loads imposed directly to the soil. Except when erected on solid rock or otherwise protected from frost, the footings shall extend below the frost line as specified in the local building code. Footings shall be permitted to be located at a depth above the frost line when protected from frost in accordance with the *Design Guide for Frost-Protected Shallow Foundations* [21]. Minimum sizes for ICF or concrete footings shall be as set forth in Table 3.1. In no case shall exterior footings be less than 12 inches (305 mm) below grade. Footings shall be supported on undisturbed natural soil or approved fill. Footings shall be stepped where it is necessary to change the elevation of the top surface of the footings. Foundations erected on soils with a bearing value of less than 2,000 psf (96 kPa) shall be designed in accordance with accepted engineering practice.

3.2 ICF Foundation Wall Requirements

The minimum wall thickness shall be greater than or equal to the wall thickness of the wall story above. A minimum of one No. 4 bar shall extend across all cold joints at a spacing not to exceed 24 inches (610 mm) on center. Cold joint reinforcement shall have a minimum of 12 inches (305 mm) embedment on both sides of all cold joints.

> **Exception:** Vertical wall reinforcement required in accordance with this section is permitted to be used in lieu of cold joint reinforcement.

Vertical wall reinforcement required in this section and interrupted by wall openings shall be placed such that one vertical bar is located within 12 inches (305 mm) of each side of the opening.

3.2.1 ICF Walls with Slab-on-Grade

ICF stem walls and monolithic slabs-on-grade shall be constructed in accordance with Figure 3.1. Vertical and horizontal wall reinforcement shall be in accordance with Section 4.0 for the above- and below-grade portions of stem walls.

3.2.2 ICF Crawlspace Walls

ICF crawlspace walls shall be constructed in accordance with Figure 3.2 and shall be laterally supported at the top and bottom of the wall in accordance with Section 6.0. A minimum of one continuous horizontal No. 4 bar shall be placed within 12 inches (305 mm) of the top of the crawlspace wall. Vertical wall reinforcement shall be the greater of that required in Table 3.2 or, if supporting an ICF wall, that required in Section 4.0 for the wall above.

3.2.3 ICF Basement Walls

ICF basement walls shall be constructed in accordance with Figure 3.3 and shall be laterally supported at the top and bottom of the wall in accordance with Section 6.0. Horizontal wall reinforcement shall be provided in accordance with Table 3.3. Vertical wall reinforcement shall be provided in accordance with Tables 3.4 through 3.9.

3.3 ICF Foundation Wall Coverings

3.3.1 Interior Covering

Rigid foam plastic on the interior of habitable spaces shall be covered with a minimum of 1/2-inch (13-mm) gypsum board or an approved finish material that provides a thermal barrier to limit the average temperature rise of the unexposed surface to no more than 250 degrees F (121 degrees C) after 15 minutes of fire exposure in accordance with ASTM E 119 [22].

The use of vapor retarders shall be in accordance with the authority having jurisdiction.

3.3.2 Exterior Covering

ICFs constructed of rigid foam plastics shall be protected from sunlight and physical damage by the application of an approved exterior covering. All ICFs shall be covered with approved materials installed to provide an adequate barrier against the weather. The use of vapor retarders and air barriers shall be in accordance with the authority having jurisdiction.

ICF foundation walls enclosing habitable or storage space shall be dampproofed from the top of the footing to the finished grade. In areas where a high water table or other severe soil-water conditions are known to exist, exterior ICF foundation walls enclosing habitable or storage space shall be waterproofed with a membrane extending from the top of the footing to the finished grade. Dampproofing and waterproofing materials for foam insulation forms shall be nonpetroleum-based and compatible with the form. Dampproofing and waterproofing materials for forms other than foam insulation shall be compatible with the form material and shall be applied in accordance with the manufacturer's recommendations.

3.4 Termite Protection Requirements

Structures consisting of materials subject to termite attack (i.e., untreated wood) shall be protected against termite infestation in accordance with the local building code. When materials susceptible to termite attack are placed on or above ICF construction, the ICF foundation walls in areas subject to termite infestation shall be protected by approved chemical soil treatment, physical barriers (i.e., termite shields), or any combination of these methods in accordance with the local building code and acceptable practice.

TABLE 3.1
MINIMUM WIDTH OF ICF AND CONCRETE FOOTINGS FOR ICF WALLS[1,2] (inches)

Maximum Number of Stories	Minimum Load-Bearing Value of Soil (psf)				
	2,000	2,500	3,000	3,500	4,000
5.5-Inch Flat, 6-Inch Waffle-Grid, or 6-Inch Screen-Grid ICF Wall Thickness[3]					
One Story	11	9	8	6	6
Two Story	20	15	13	11	10
Three Story[4]	31	23	18	15	13
7.5-Inch Flat or 8-Inch Waffle-Grid ICF Wall Thickness[3]					
One Story	13	10	9	8	8
Two Story	22	18	15	13	11
Three Story[5]	36	27	21	18	15
9.5-Inch Flat ICF Wall Thickness[3]					
One Story	15	12	10	10	10
Two Story	26	21	17	15	13
Three Story	41	31	24	20	17

For SI: 1 foot = 0.3048 m; 1 inch = 25.4 mm; 1 psf = 47.8804 Pa

1 Minimum footing thickness shall be the greater of 11 inches (279 mm) when a dowel is required in accordance with Section 6.0, one-third of the footing width, or 6 inches (152 mm).

2 Required minimum footing widths that are greater than the ICF wall thickness shall be increased as required to allow for a minimum 2-inch (51-mm) footing width projection on both sides of the ICF wall. The footing width projection shall be measured from the face of the concrete in the ICF wall to the edge of the footing.

3 Actual thickness is shown for flat walls while nominal thickness is given for waffle- and screen-grid walls. Refer to Section 2.0 for actual waffle- and screen-grid thickness and dimensions.

4 Applicable also for 7.5-inch- (191-mm-) thick or 9.5-inch- (241-mm-) thick flat ICF story supporting two 3.5-inch-thick flat ICF stories.

5 Applicable also for 9.5-inch- (241-mm-) thick flat ICF wall story supporting two 5.5-inch- (140-mm-) thick flat ICF stories.

TABLE 3.2
MINIMUM VERTICAL WALL REINFORCEMENT FOR ICF CRAWLSPACE WALLS [1,2,3,4]

Shape of Concrete Walls	Wall Thickness[5] (inches)	Maximum Unbalanced Backfill Height[6] (feet)	Minimum Vertical Reinforcement		
			Maximum Equivalent Fluid Density 30 pcf	Maximum Equivalent Fluid Density 45 pcf	Maximum Equivalent Fluid Density 60 pcf
Flat	3.5 [7]	4	#4@32"	#3@18"; #4@28"; #5@38"	#3@12"; #4@22"; #5@28"
	5.5	4	#4@48"	#4@48"	#4@48"
	7.5	4	N/R	N/R	N/R
Waffle-Grid	6	4	#4@48"	#4@48"	#3@12"; #4@24"; #5@36"
	8	4	N/R	N/R	N/R
Screen-Grid	6	4	#4@48"	#4@48"	#3@12"; #4@24"; #5@36"

For SI: 1 foot = 0.3048 m; 1 inch = 25.4 mm; 1 pcf = 16.0179 kg/m^3

[1] N/R indicates no vertical wall reinforcement is required.
[2] Applicable only to crawlspace walls 5 feet (1.5 m) or less in height.
[3] Interpolation shall not be permitted.
[4] Walls shall be laterally supported at the top before backfilling.
[5] Actual thickness is shown for flat walls while nominal thickness is given for waffle- and screen-grid walls. Refer to Section 2.0 for actual waffle- and screen-grid thickness and dimensions.
[6] Refer to Section 1.0 for the definition of unbalanced backfill height.
[7] Applicable only to one-story construction with floor bearing on top of crawlspace wall.

TABLE 3.3
MINIMUM HORIZONTAL WALL REINFORCEMENT FOR ICF BASEMENT WALLS

Maximum Height of Basement Wall feet (meters)	Location of Horizontal Reinforcement
8 (2.4)	One No. 4 bar within 12 inches (305 mm) of the top of the wall story and one No. 4 bar near mid-height of the wall story
9 (2.7)	One No. 4 bar within 12 inches (305 mm) of the top of the wall story and one No. 4 bar near third points in the wall story
10 (3.0)	One No. 4 bar within 12 inches (305 mm) of the top of the wall story and one No. 4 bar near third points in the wall story

TABLE 3.4
MINIMUM VERTICAL WALL REINFORCEMENT FOR 5.5-INCH- (140-MM-) THICK FLAT ICF BASEMENT WALLS [1,2,3]

Max. Height of Basement Wall (feet)	Maximum Unbalanced Backfill Height[4] (feet)	Minimum Vertical Reinforcement		
		Maximum Equivalent Fluid Density 30 pcf	Maximum Equivalent Fluid Density 45 pcf	Maximum Equivalent Fluid Density 60 pcf
8	4	#4@48"	#4@48"	#4@48"
	5	#4@48"	#3@12"; #4@22"; #5@32"; #6@40"	#3@8"; #4@14"; #5@20"; #6@26"
	6	#3@12"; #4@22"; #5@30"; #6@40"	#3@8"; #4@14"; #5@20"; #6@24"	#3@6"; #4@10"; #5@14"; #6@20"
	7	#3@8"; #4@14"; #5@22"; #6@26"	#3@5"; #4@10"; #5@14"; #6@18"	#3@4"; #4@6"; #5@10"; #6@14"
9	4	#4@48"	#4@48"	#4@48"
	5	#4@48"	#3@12"; #4@20"; #5@28"; #6@36"	#3@8"; #4@14"; #5@20"; #6@22"
	6	#3@10"; #4@20"; #5@28"; #6@34"	#3@6"; #4@12"; #5@18"; #6@20"	#4@8"; #5@14"; #6@16"
	7	#3@8"; #4@14"; #5@20"; #6@22"	#4@8"; #5@12"; #6@16"	#4@6"; #5@10"; #6@12"
	8	#3@6"; #4@10"; #5@14"; #6@16"	#4@6"; #5@10"; #6@12"	#4@4"; #5@6"; #6@8"
10	4	#4@48"	#4@48"	#4@48"
	5	#4@48"	#3@10"; #4@18"; #5@26"; #6@30"	#3@6"; #4@14"; #5@18"; #6@20"
	6	#3@10"; #4@18"; #5@24"; #6@30"	#3@6"; #4@12"; #5@16"; #6@18"	#3@4"; #4@8"; #5@12"; #6@14"
	7	#3@6"; #4@12"; #5@16"; #6@18"	#3@4"; #4@8"; #5@12	#4@6"; #5@8"; #6@10"
	8	#3@4"; #4@8"; #5@12"; #6@14"	#4@6"; #5@8"; #6@12"	#4@4"; #5@6"; #6@8"
	9	#3@4"; #4@6"; #5@10"; #6@12"	#4@4"; #5@6"; #6@8"	#5@4"; #6@6"

For SI: 1 foot = 0.3048 m; 1 inch = 25.4 mm; 1 pcf = 16.0179 kg/m^3

1 Deflection criterion is *L/240*, where *L* is the height of the basement wall in inches.
2 Interpolation shall not be permitted.
3 Walls shall be laterally supported at the top before backfilling.
4 Refer to Section 1.0 for the definition of unbalanced backfill height.

TABLE 3.5
MINIMUM VERTICAL WALL REINFORCEMENT FOR 7.5-INCH- (191-MM-) THICK FLAT ICF BASEMENT WALLS [1,2,3,4]

Max. Height of Basement Wall (feet)	Maximum Unbalanced Backfill Height[5] (feet)	Minimum Vertical Reinforcement		
		Maximum Equivalent Fluid Density 30 pcf	Maximum Equivalent Fluid Density 45 pcf	Maximum Equivalent Fluid Density 60 pcf
8	4	N/R	N/R	N/R
	5	N/R	N/R	N/R
	6	N/R	N/R	N/R
	7	N/R	#3@8"; #4@14"; #5@20"; #6@28"	#3@6"; #4@10"; #5@16"; #6@20"
9	4	N/R	N/R	N/R
	5	N/R	N/R	N/R
	6	N/R	N/R	#3@8"; #4@14"; #5@20"; #6@28"
	7	N/R	#3@6"; #4@12"; #5@18"; #6@26"	#3@4"; #4@8"; #5@14"; #6@18"
	8	#3@8"; #4@14"; #5@22"; #6@28"	#3@4"; #4@8"; #5@14"; #6@18"	#3@4"; #4@6"; #5@10"; #6@14"
10	4	N/R	N/R	N/R
	5	N/R	N/R	N/R
	6	N/R	N/R	#3@6"; #4@12"; #5@18"; #6@26"
	7	N/R	#3@6"; #4@12"; #5@18"; #6@24"	#3@4"; #4@8"; #5@12"; #6@18"
	8	#3@6"; #4@12"; #5@20"; #6@26"	#3@4"; #4@8"; #5@12"; #6@16"	#3@4"; #4@6"; #5@8"; #6@12"
	9	#3@6"; #4@10"; #5@14"; #6@20"	#3@4"; #4@6"; #5@10"; #6@12"	#4@4"; #5@6"; #6@10"

For SI: 1 foot = 0.3048 m; 1 inch = 25.4 mm; 1 pcf = 16.0179 kg/m^3

1 N/R indicates no reinforcement is required.
2 Deflection criterion is *L/240*, where *L* is the height of the basement wall in inches.
3 Interpolation shall not be permitted.
4 Walls shall be laterally supported at the top before backfilling.
5 Refer to Section 1.0 for the definition of unbalanced backfill height.

TABLE 3.6
MINIMUM VERTICAL WALL REINFORCEMENT FOR 9.5-INCH- (241-MM-) THICK FLAT ICF BASEMENT WALLS [1,2,3,4]

Max. Height of Basement Wall (feet)	Maximum Unbalanced Backfill Height[5] (feet)	Minimum Vertical Reinforcement		
		Maximum Equivalent Fluid Density 30 pcf	Maximum Equivalent Fluid Density 45 pcf	Maximum Equivalent Fluid Density 60 pcf
8	4	N/R	N/R	N/R
	5	N/R	N/R	N/R
	6	N/R	N/R	N/R
	7	N/R	N/R	N/R
9	4	N/R	N/R	N/R
	5	N/R	N/R	N/R
	6	N/R	N/R	N/R
	7	N/R	N/R	#3@6"; #4@12"; #5@18"; #6@26"
	8	N/R	#3@6"; #4@12"; #5@18"; #6@26"	#3@4"; #4@8"; #5@14"; #6@18"
10	4	N/R	N/R	N/R
	5	N/R	N/R	N/R
	6	N/R	N/R	#3@10"; #4@18"; #5@26"; #6@36"
	7	N/R	N/R	#3@6"; #4@10"; #5@18"; #6@24"
	8	N/R	#3@6"; #4@12"; #5@16"; #6@24"	#3@4"; #4@8"; #5@12"; #6@16"
	9	N/R	#3@4"; #4@8"; #5@12"; #6@18"	#3@4"; #4@6"; #5@10"; #6@12"

For SI: 1 foot = 0.3048 m; 1 inch = 25.4 mm; 1 pcf = 16.0179 kg/m^3

1 N/R indicates no vertical wall reinforcement is required.
2 Deflection criterion is *L/240*, where *L* is the height of the basement wall in inches.
3 Interpolation shall not be permitted.
4 Walls shall be laterally supported at the top before backfilling.
5 Refer to Section 1.0 for the definition of unbalanced backfill height.

TABLE 3.7
MINIMUM VERTICAL WALL REINFORCEMENT FOR 6-INCH (152-MM) WAFFLE-GRID ICF BASEMENT WALLS[1,2,3,4,5]

Max. Height of Basement Wall (feet)	Maximum Unbalanced Backfill Height[6] (feet)	Minimum Vertical Reinforcement		
		Maximum Equivalent Fluid Density 30 pcf	Maximum Equivalent Fluid Density 45 pcf	Maximum Equivalent Fluid Density 60 pcf
8	4	#4@48”	#3@12”; #4@24”; #5@24”	#3@12”; #4@12”
	5	#3@12”; #4@12”; #5@24”	#4@12”; #5@12”	#7@12”
	6	#4@12”; #5@12”	#7@12”	#7@12”
	7	#6@12”	Design Required	Design Required
9	4	#4@48”	#3@12”; #4@12”; #5@24”	#3@12”; #4@12”
	5	#3@12”; #4@12”	#4@12”; #5@12”	#7@12”
	6	#5@12”; #6@12”	Design Required	Design Required
	7	Design Required	Design Required	Design Required
	8	Design Required	Design Required	Design Required
10	4	#4@48”	#4@12”; #5@12”	#5@12”; #6@12”
	5	#3@12”; #4@12”	#7@12”	Design Required
	6	#7@12”	Design Required	Design Required
	7	Design Required	Design Required	Design Required
	8	Design Required	Design Required	Design Required
	9	Design Required	Design Required	Design Required

For SI: 1 foot = 0.3048 m; 1 inch = 25.4 mm; 1 pcf = 16.0179 kg/m^3

[1] N/R indicates no vertical wall reinforcement is required.
[2] Deflection criterion is *L/240*, where *L* is the height of the basement wall in inches.
[3] Interpolation shall not be permitted.
[4] Nominal thickness is given; refer to Section 2.0 for actual wall thickness.
[5] Walls shall be laterally supported at the top before backfilling.
[6] Refer to Section 1.0 for the definition of unbalanced backfill height.

TABLE 3.8
MINIMUM VERTICAL WALL REINFORCEMENT FOR 8-INCH (203-MM) WAFFLE-GRID ICF BASEMENT WALLS[1,2,3,4,5]

Max. Height of Basement Wall (feet)	Maximum Unbalanced Backfill Height[6] (feet)	Minimum Vertical Reinforcement		
		Maximum Equivalent Fluid Density 30 pcf	Maximum Equivalent Fluid Density 45 pcf	Maximum Equivalent Fluid Density 60 pcf
8	4	N/R	N/R	N/R
	5	N/R	#3@12"; #4@24"; #5@36"	#3@12"; #4@12"; #5@24"
	6	#3@12"; #4@24"; #5@36"	#4@12"; #5@24"	#4@12"; #5@12"
	7	#3@12"; #4@12"; #5@12"; #6@24"	#4@12"; #5@12"	#5@12"; #6@12"
9	4	N/R	N/R	N/R
	5	N/R	#3@12"; #4@12"; #5@24"	#3@12"; #4@12"; #5@24"
	6	#3@12"; #4@24"; #5@24"	#4@12"; #5@12"	#4@12"; #5@12"
	7	#4@12"; #5@24"; #6@24"	#5@12"; #6@12"	#5@12"; #6@12"
	8	#4@12"; #5@12"	#5@12"; #6@12"	#8@12"
10	4	N/R	#3@12"; #4@24"; #5@24"; #6@36"	#3@12"; #4@12"; #5@24"
	5	N/R	#3@12"; #4@24"; #5@24"; #6@36"	#4@12"; #5@24"
	6	#3@12"; #4@12" #5@24"	#4@12"; #5@12"	#5@12"; #6@12"
	7	#4@12"; #5@12"	#5@12"; #6@12"	#6@12"; #7@12"
	8	#4@12"; #5@12"	#6@12"; #7@12"	#8@12"
	9	#5@12"; #6@12"	#8@12"	#8@12"

For SI: 1 foot = 0.3048 m; 1 inch = 25.4 mm; 1 pcf = 16.0179 kg/m^3

1 N/R indicates no vertical wall reinforcement is required.
2 Deflection criterion is *L/240*, where *L* is the height of the basement wall in inches.
3 Interpolation shall not be permitted.
4 Nominal thickness is given; refer to Section 2.0 for actual wall thickness.
5 Walls shall be laterally supported at the top before backfilling.
6 Refer to Section 1.0 for the definition of unbalanced backfill height.

TABLE 3.9
MINIMUM VERTICAL WALL REINFORCEMENT FOR 6-INCH (152-MM) SCREEN-GRID ICF BASEMENT WALLS[1,2,3,4]

Max. Height of Basement Wall (feet)	Maximum Unbalanced Backfill Height[5] (feet)	Minimum Vertical Reinforcement		
		Maximum Equivalent Fluid Density 30 pcf	Maximum Equivalent Fluid Density 45 pcf	Maximum Equivalent Fluid Density 60 pcf
8	4	#4@48"	#3@12"; #4@24"; #5@36"	#3@12"; #4@12"; #5@24"
	5	#3@12"; #4@24"; #5@24"	#3@12"; #4@12"	#4@12"; #5@12"
	6	#4@12"; #5@12"	#5@12"; #6@12"	#7@12"
	7	#4@12"; #5@12"	#7@12"	Design Required
9	4	#4@48"	#3@12"; #4@24"; #5@24"	#3@12"; #4@12"; #5@12"; #6@24"
	5	#3@12"; #4@12"; #5@24"	#4@12"; #5@12"	#7@12"
	6	#4@12"; #5@12"	#7@12"	#7@12"
	7	#7@12"	Design Required	Design Required
	8	#7@12"	Design Required	Design Required
10	4	#4@48"	#3@12"; #4@12"; #5@24"; #6@24"	#3@12"; #4@12"
	5	#3@12"; #4@12"	#4@12"; #5@12"	#7@12"
	6	#4@12"; #5@12"	Design Required	Design Required
	7	#7@12"	Design Required	Design Required
	8	Design Required	Design Required	Design Required
	9	Design Required	Design Required	Design Required

For SI: 1 foot = 0.3048 m; 1 inch = 25.4 mm; 1 pcf = 16.0179 kg/m^3

1 Deflection criterion is *L/240*, where *L* is the height of the basement wall in inches.
2 Interpolation shall not be permitted.
3 Nominal thickness is given; refer to Section 2.0 for actual concrete wall thickness.
4 Walls shall be laterally supported at the top before backfilling.
5 Refer to Section 1.0 for definition of unbalanced backfill height.

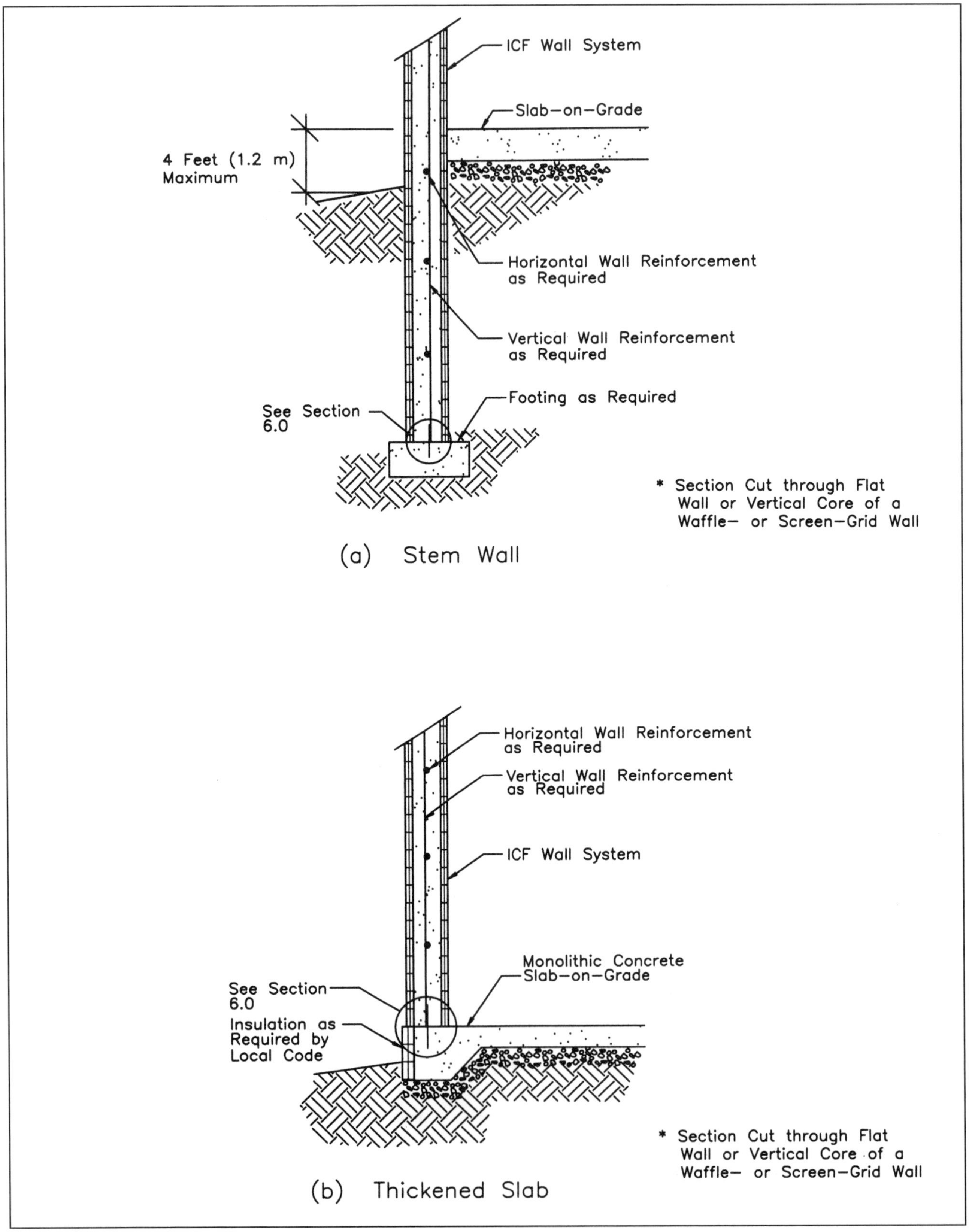

Figure 3.1 ICF Stem Wall and Monolithic Slab-on-Grade Construction

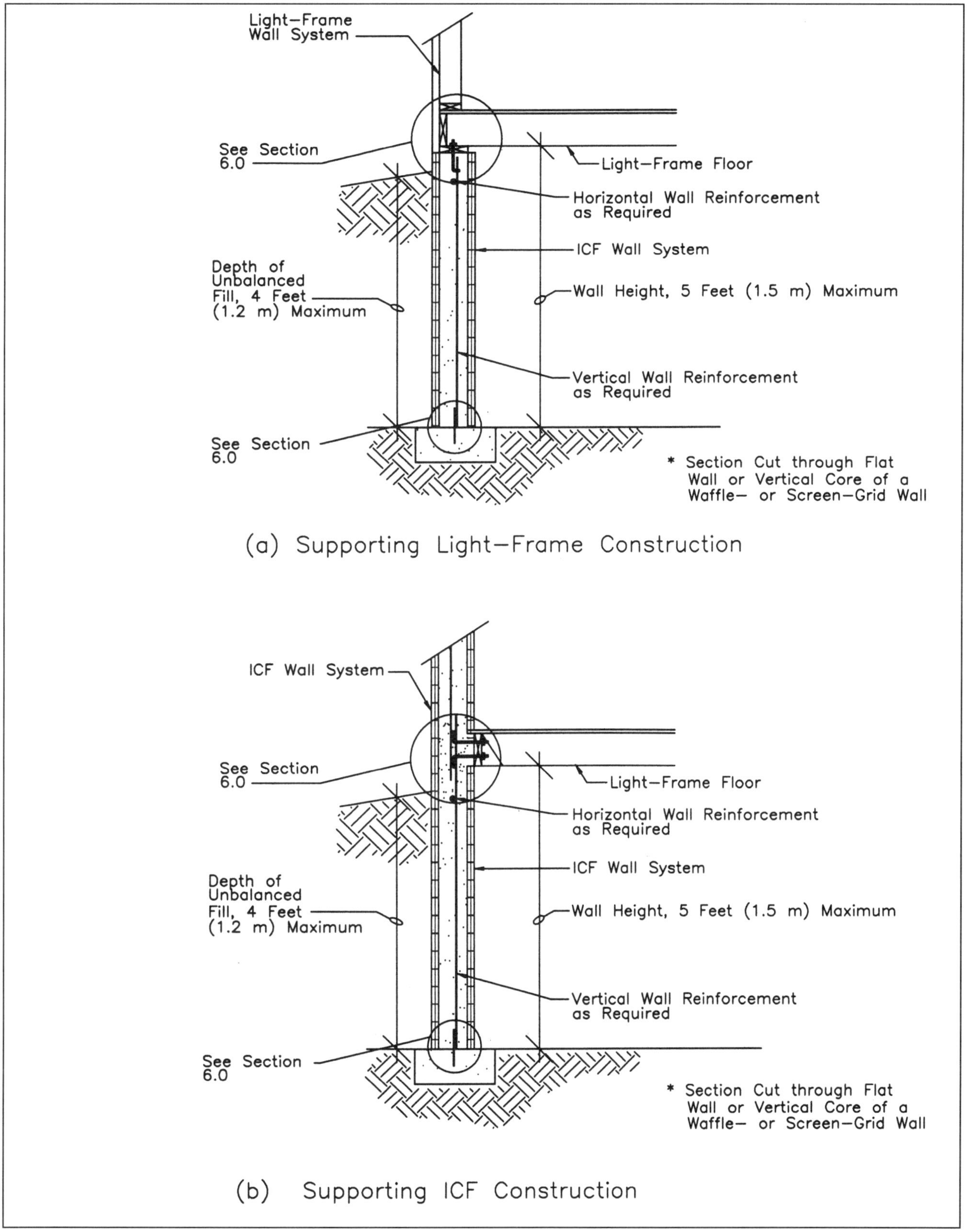

Figure 3.2 ICF Crawlspace Wall Construction

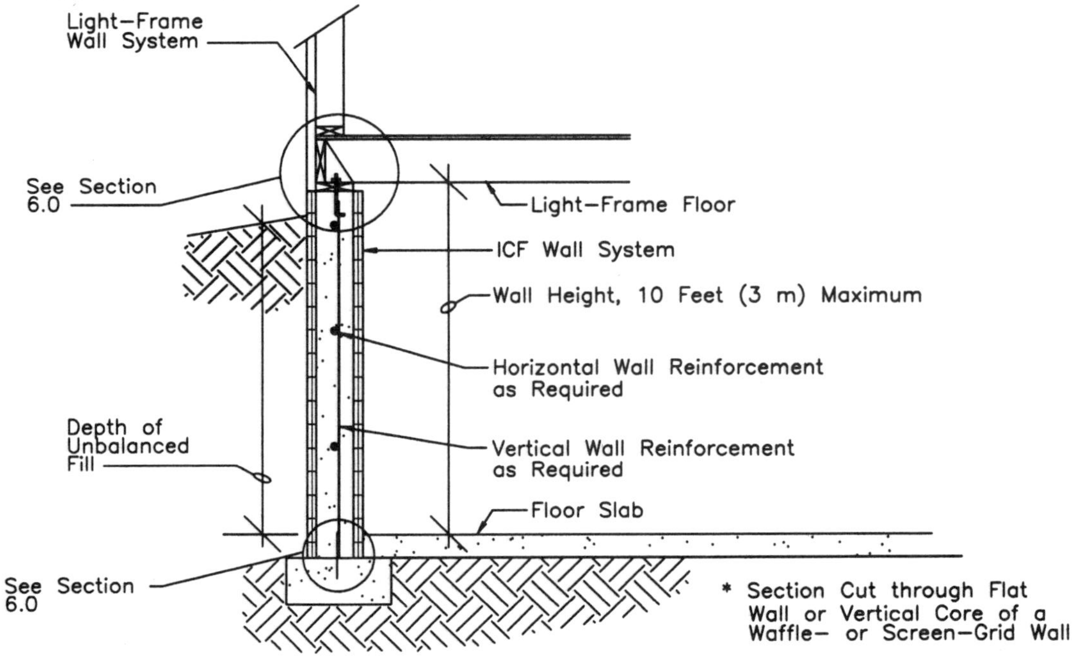

(a) Supporting Light–Frame Construction

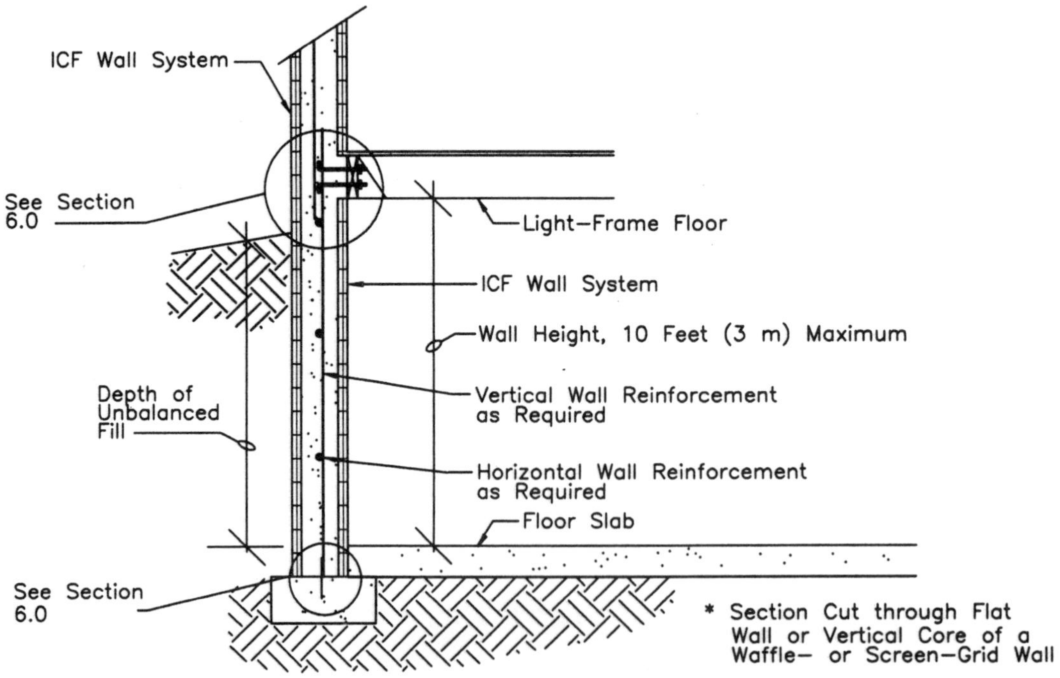

(b) Supporting ICF Construction

Figure 3.3 ICF Basement Wall Construction

4.0 ICF ABOVE-GRADE WALLS

4.1 ICF Above-Grade Wall Requirements

ICF above-grade walls shall be constructed in accordance with Figures 4.1, 4.2, or 4.3 and this section. Lateral support for above-grade ICF walls shall be provided by the roof and floor framing systems in accordance with Section 6.0.

Horizontal wall reinforcement shall be provided in accordance with Table 4.1. Vertical wall reinforcement shall be provided in accordance with Tables 4.2 through 4.4.

The minimum wall thickness shall be greater than or equal to the wall thickness of the wall story above. A minimum of one No. 4 bar shall extend across all cold joints at a spacing not to exceed 24 inches (610 mm) on center. Cold joint reinforcement shall have a minimum of 12 inches (305 mm) embedment on both sides of all cold joints.

Exception: Vertical wall reinforcement required in accordance with this section is permitted to be used in lieu of cold joint reinforcement.

Vertical wall reinforcement required in this section and interrupted by wall openings shall be placed such that one vertical bar is located within 12 inches (305 mm) of each side of the opening.

In conditions with wind speeds greater than 90 mph (145 km/hr), all vertical wall reinforcement in the top-most ICF story shall be terminated with a minimum 90 degree bend. The bend shall result in a minimum length of 24 inches (610 mm) parallel to the horizontal wall reinforcement and lie within 4 inches (102 mm) of the top surface of the ICF wall near the roof level. In addition, all horizontal wall reinforcement at building corners shall be terminated with a bend resulting in a minimum length of 24 inches (610 mm) embedment in the intersecting wall.

Exception: A standard 90 degree hook is permitted to be used in lieu of bending the vertical or horizontal wall reinforcement provided that a minimum 24-inch (610-mm) lap splice is provided.

4.2 ICF Above-Grade Wall Coverings

4.2.1 Interior Covering

Rigid foam plastic on the interior of habitable spaces shall be covered with a minimum of 1/2-inch (13-mm) gypsum board or an approved finish material that provides a thermal barrier to limit the average temperature rise of the unexposed surface to no more than 250 degrees F (121 degrees C) after 15 minutes of fire exposure in accordance with ASTM E 119 [22]. The use of vapor retarders and air barriers shall be in accordance with the authority having jurisdiction.

4.2.2 Exterior Covering

ICFs constructed of rigid foam plastics shall be protected from sunlight and physical damage by the application of an approved exterior covering. All ICFs shall be covered with approved materials installed to provide a barrier against the weather. Use of air barriers and vapor retarders shall be in accordance with the authority having jurisdiction.

TABLE 4.1
MINIMUM HORIZONTAL WALL REINFORCEMENT FOR ICF ABOVE-GRADE WALLS

ICF Wall Type and Minimum Wall Thickness inches (mm)[1]	Maximum Height of Wall Story feet (meters)	Location of Horizontal Reinforcement
Flat 3.5 (89)	8 (2.4)	One No. 4 bar within 12 inches (305 mm) of the top of the wall story and one No. 4 bar near third points in the wall story
	9 (2.7)	One No. 4 bar within 12 inches (305 mm) of the top of the wall story and one No. 4 bar near third points in the wall story
	10 (3.0)	One No. 4 bar within 12 inches (305 mm) of the top of the wall story and one No. 4 bar near fourth points in the wall story
Flat 5.5 (140) or Waffle-Grid 6 (152) or Screen-Grid 6 (152)	8 (2.4)	One No. 4 bar within 12 inches (305 mm) of the top of the wall story and one No. 4 bar near mid-height of the wall story
	9 (2.7)	One No. 4 bar within 12 inches (305 mm) of the top of the wall story and one No. 4 bar near third points in the wall story
	10 (3.0)	One No. 4 bar within 12 inches (305 mm) of the top of the wall story and one No. 4 bar near third points in the wall story

[1] Actual thickness is shown for flat walls while nominal thickness is given for waffle- and screen-grid walls. Refer to Section 2.0 for actual waffle- and screen-grid thickness and dimensions.

TABLE 4.2
MINIMUM VERTICAL WALL REINFORCEMENT FOR FLAT ICF ABOVE-GRADE WALLS [1,2,3]

Max. Wind Speed (mph)	Maximum Wall Height per Story (feet)	Minimum Vertical Reinforcement					
		Supporting Light-Frame Roof Only		Supporting Light-Frame Second Story and Roof		Supporting ICF Second Story and Light-Frame Roof	
		Minimum Wall Thickness (inches)					
		3.5	5.5	3.5	5.5	3.5	5.5
70	8	#4@32"	N/R	#4@32"	N/R	#4@32"	N/R
	9	#4@32"	N/R	#4@32"	N/R	#3@20"; #4@24"; #5@26"	N/R
	10	#4@32"	N/R	#4@32"	N/R	#3@20"; #4@24"; #5@26"	N/R
80	8	#4@32"	N/R	#4@32"	N/R	#4@32"	N/R
	9	#4@32"	N/R	#3@28"; #4@32"; #5@34"	N/R	#3@14"; #4@18"; #5@20"	N/R
	10	#3@16"; #4@26"; #5@34"	N/R	#3@16"; #4@26"; #5@34"	N/R	Design Required	N/R
90	8	#4@32"	#4@96"	#4@32"	#4@96"	#4@32"	#4@96"
	9	#3@16"; #4@26"; #5@34"	#4@96"	#3@18"; #4@20"; #5@22"	#4@96"	Design Required	#4@96"
	10	Design Required	#4@96"	#3@10"; #4@12"; #5@14"	#4@96"	Design Required	#4@96"
100	8	#3@18"; #4@30"; #5@40"	#4@96"	#3@18"; #4@30"; #5@40"	#4@96"	#4@32"	#4@96"
	9	#3@12"; #4@22"; #5@30"	#4@96"	#3@12"; #4@16"; #5@16"	#4@96"	Design Required	#4@96"
	10	Design Required	#4@96"	Design Required	#4@96"	Design Required	#4@96"
110	8	#3@14"; #4@20"; #5@24"	#4@96"	#3@14"; #4@18"; #5@20"	#4@96"	Design Required	#4@96"
	9	Design Required	#4@96"	Design Required	#4@96"	Design Required	#4@96"
	10	Design Required	#4@96"	Design Required	#4@96"	Design Required	#4@96"

For SI: 1 foot = 0.3048 m; 1 inch = 25.4 mm; 1 mph = 1.6093 km/hr

[1] N/R indicates no vertical wall reinforcement is required.
[2] Deflection criterion is *L/240*, where *L* is the height of the wall story in inches.
[3] Interpolation shall not be permitted.

TABLE 4.3
MINIMUM VERTICAL WALL REINFORCEMENT FOR WAFFLE-GRID ICF ABOVE-GRADE WALLS [1,2,3]

Max. Wind Speed	Maximum Wall Height per Story	Minimum Vertical Reinforcement					
		Supporting Light-Frame Roof Only		Supporting Light-Frame Second Story and Roof		Supporting ICF Second Story and Light-Frame Roof	
		Minimum Wall Thickness [4] (inches)					
(mph)	(feet)	6	8	6	8	6	8
70	8	N/R	N/R	N/R	N/R	N/R	N/R
	9	N/R	N/R	N/R	N/R	N/R	N/R
	10	N/R	N/R	N/R	N/R	N/R	N/R
80	8	N/R	N/R	N/R	N/R	N/R	N/R
	9	N/R	N/R	N/R	N/R	N/R	N/R
	10	N/R	N/R	N/R	N/R	N/R	N/R
90	8	#4@96"	#4@96"	#4@96"	#4@96"	#4@96"	#4@96"
	9	#4@96"	#4@96"	#4@96"	#4@96"	#4@96"	#4@96"
	10	#3@12"; #4@24"; #5@36"	#4@96"	#3@12"; #4@24"; #5@24"	#4@96"	#3@12"; #4@24"; #5@24"	#4@96"
100	8	#4@96"	#4@96"	#4@96"	#4@96"	#4@96"	#4@96"
	9	#3@12"; #4@24"; #5@36"	#4@96"	#3@12"; #4@24"; #5@36"	#4@96"	#3@12"; #4@24"; #5@24"	#4@96"
	10	#3@12"; #4@24"; #5@24"	#4@96"	#3@12"; #4@24"; #5@24"	#4@96"	#3@12"; #4@12"	#4@96"
110	8	#3@12"; #4@24"; #5@36"	#4@96"	#3@12"; #4@24"; #5@36"	#4@96"	#3@12"; #4@24"; #5@24"	#4@96"
	9	#3@12"; #4@24"; #5@24"	#4@96"	#3@12"; #4@24"; #5@24"	#4@96"	#3@12"; #4@12"	#4@96"
	10	#3@12"; #4@12"; #5@24"	#4@96"	Design Required	#4@96"	Design Required	#4@96"

For SI: 1 foot = 0.3048 m; 1 inch = 25.4 mm; 1 mph = 1.6093 km/hr

1 N/R indicates no vertical wall reinforcement is required.
2 Deflection criterion is *L/240*, where *L* is the height of the wall story in inches.
3 Interpolation shall not be permitted.
4 Nominal thickness is given; refer to Section 2.0 for actual concrete wall thickness.

TABLE 4.4
MINIMUM VERTICAL WALL REINFORCEMENT FOR SCREEN-GRID ICF ABOVE-GRADE WALLS [1,2,3]

Max. Wind Speed (mph)	Maximum Wall Height per Story (feet)	Minimum Vertical Reinforcement		
		Supporting Light-Frame Roof Only	Supporting Light-Frame Second Story and Roof	Supporting ICF Second Story and Light-Frame Roof
		Minimum Wall Thickness [4] (inches)		
		6	6	6
70	8	N/R	N/R	N/R
	9	N/R	N/R	N/R
	10	N/R	N/R	N/R
80	8	N/R	N/R	N/R
	9	N/R	N/R	N/R
	10	N/R	N/R	N/R
90	8	#4@96"	#4@96"	#4@96"
	9	#4@96"	#4@96"	#4@96"
	10	#3@12"; #4@24"; #5@36"	#3@12"; #4@24"; #5@36"	#4@96"
100	8	#4@96"	#4@96"	#4@96"
	9	#3@24"; #4@36"	#3@24"; #4@36"; #5@36"	#4@96"
	10	#3@12"; #4@24"; #5@36"	#4@24"; #5@24"	#3@12"; #4@24"; #5@24"
110	8	#3@24"; #4@36"; #5@48"	#3@24"; #4@36"; #5@48"	#4@96"
	9	#3@12"; #4@24"; #5@36"	#3@12"; #4@24"; #5@36"	#3@12"; #4@24"; #5@24"
	10	#3@12"; #4@12"	#4@24"; #5@24"	#3@12"; #4@12"; #5@24"

For SI: 1 foot = 0.3048 m; 1 inch = 25.4 mm; 1 mph = 1.6093 km/hr

1 N/R indicates no vertical wall reinforcement is required.
2 Deflection criterion is *L/240*, where *L* is the height of the wall story in inches.
3 Interpolation shall not be permitted.
4 Nominal thickness is given; refer to Section 2.0 for actual concrete wall thickness.

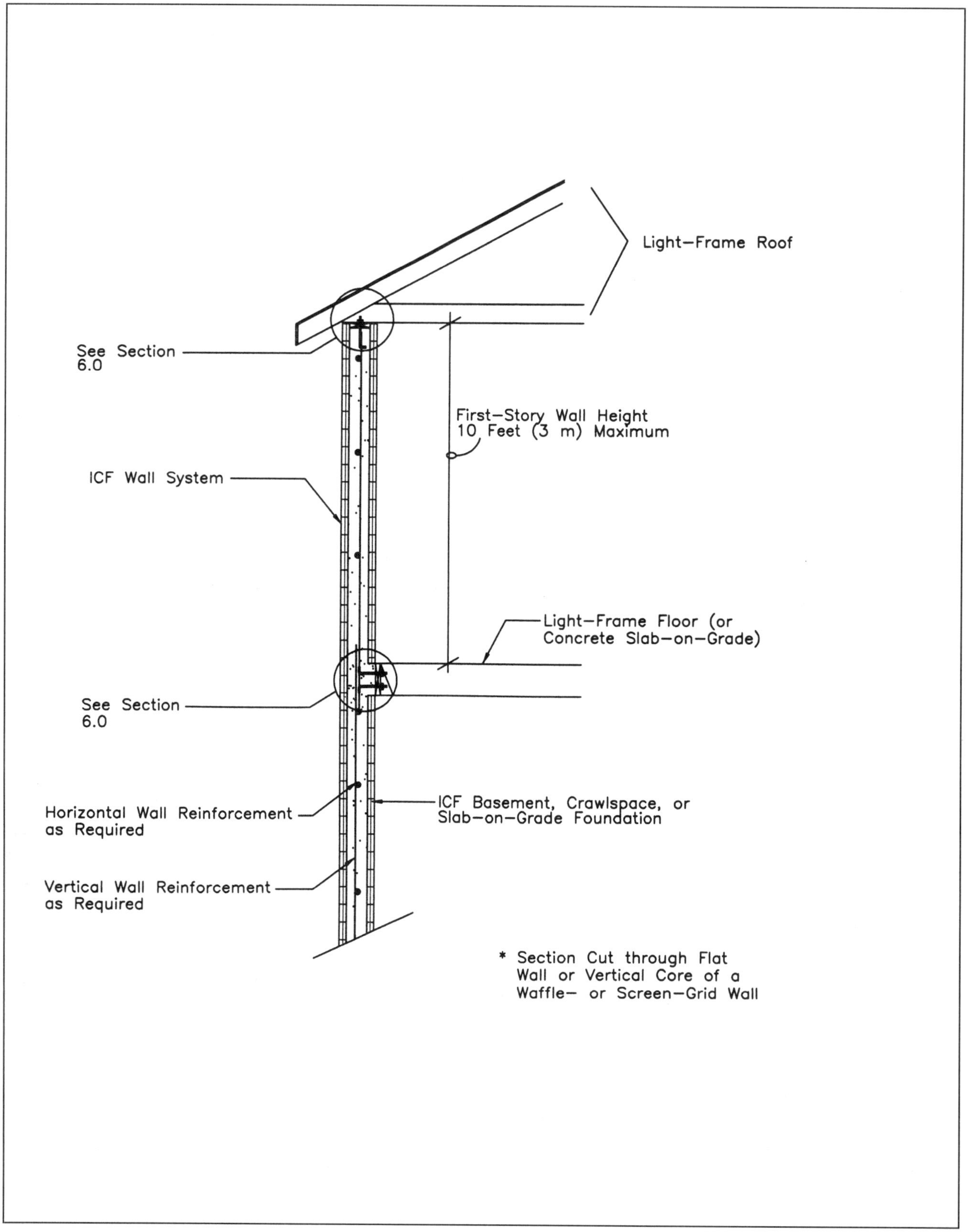

Figure 4.1 ICF Wall Supporting Light-Frame Roof

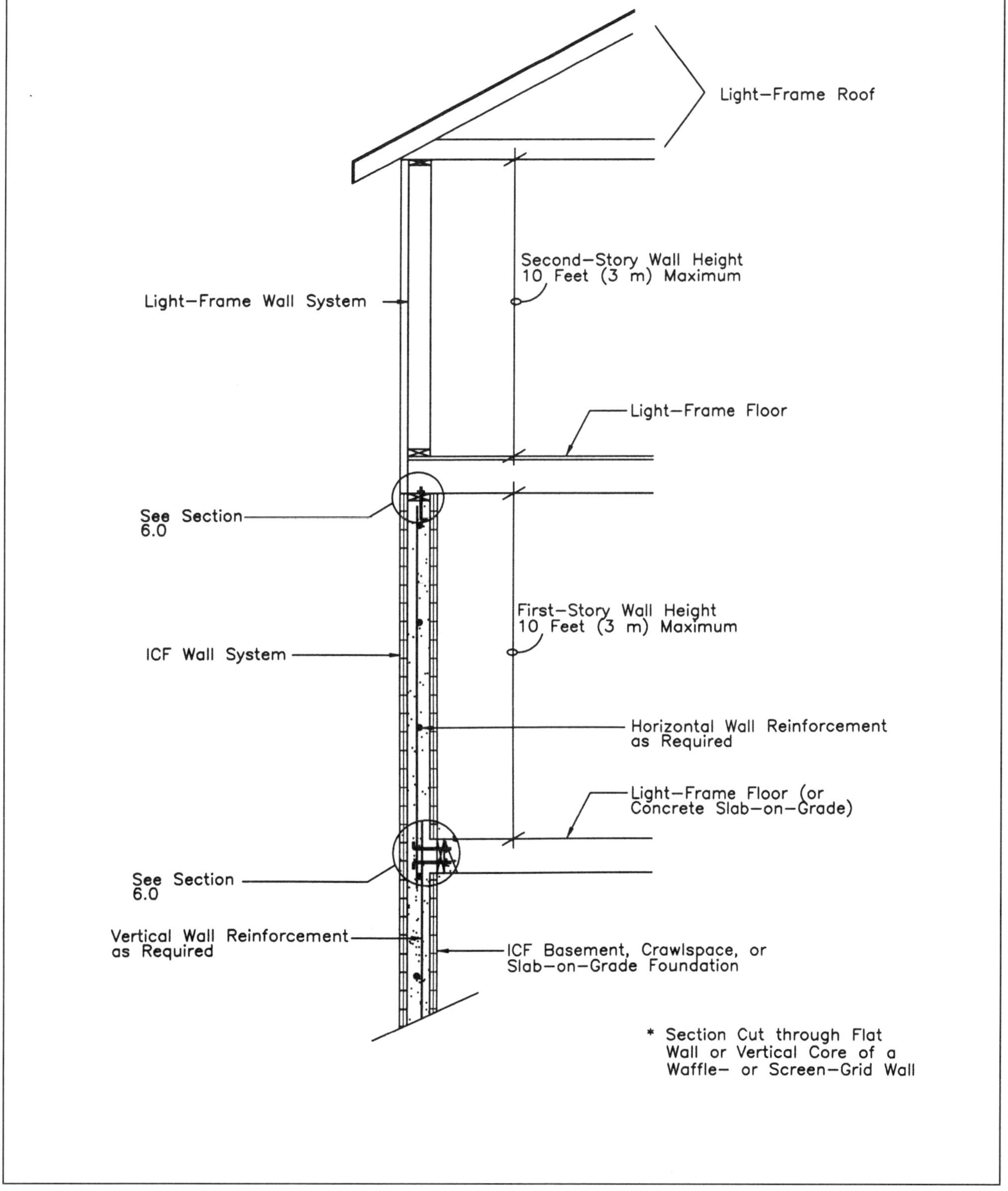

Figure 4.2 ICF Wall Supporting Light-Frame Second Story and Roof

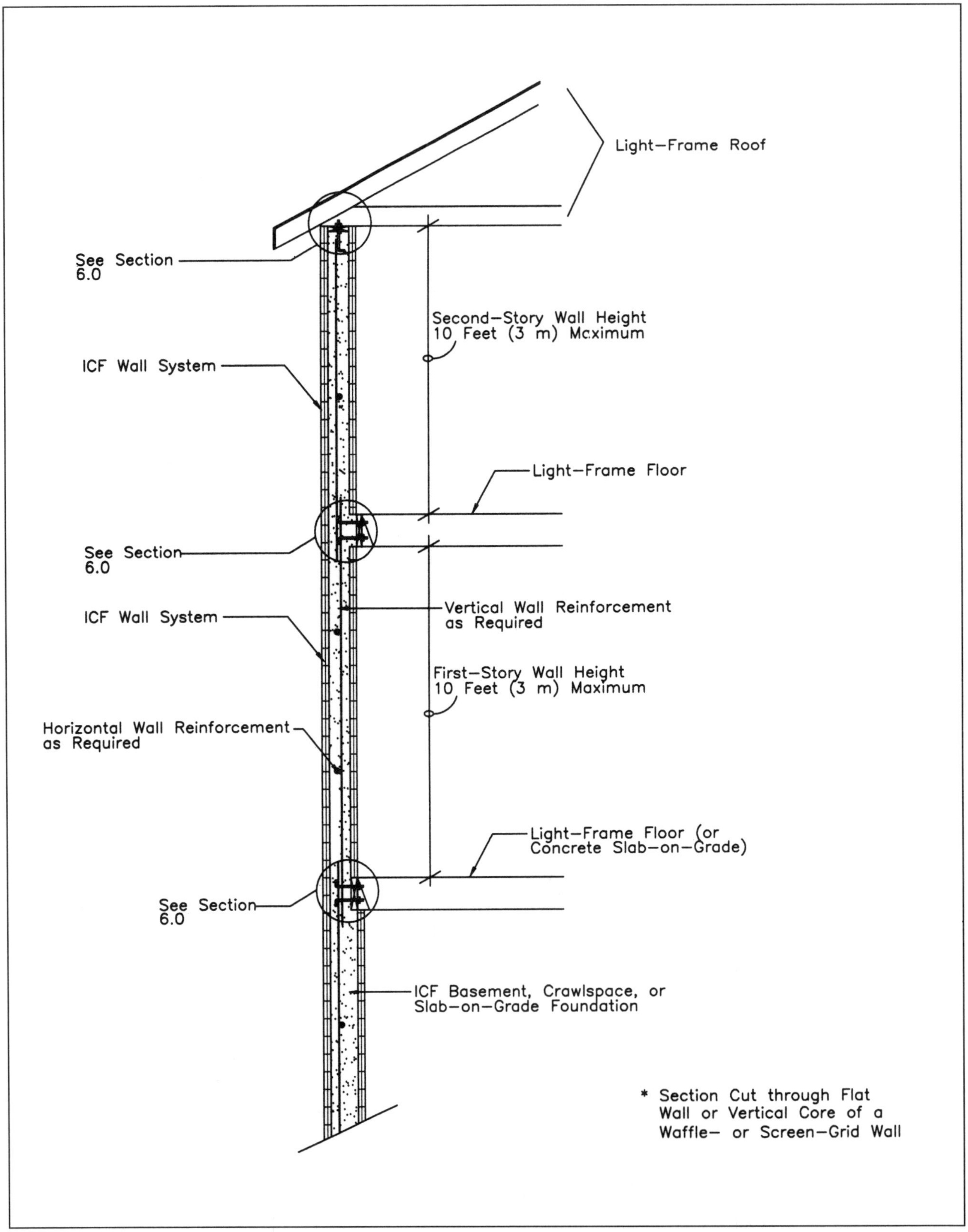

Figure 4.3 ICF Wall Supporting ICF Second Story and Light-Frame Roof

5.0 ICF WALL OPENING REQUIREMENTS

5.1 Minimum Length of ICF Wall without Openings

Values for adjustment factors are provided in Table 5.1 for use with Table 5.2 when determining the minimum percentage of solid wall length for exterior ICF walls. Exterior ICF walls shall have a minimum percentage of solid wall length for exterior ICF walls in accordance with this section, Table 5.2, and Figure 5.1. Values for base percentages of solid wall length are provided in Table 5.3 for use with Table 5.2. The minimum and base percentage of solid wall length in Tables 5.2 and 5.3 shall include only those solid wall segments that are a minimum of 24 inches (610 mm) in length. The maximum allowable spacing of wall segments at least 24 inches (610 mm) in length shall be 18 feet (5.5 m) on center. A minimum length of 24 inches (610 mm) of solid wall segment, extending the full height of each wall story, shall occur at all interior and exterior corners of exterior walls.

5.2 Reinforcement around Openings

Openings in ICF walls shall be reinforced in accordance with this section. Wall openings shall have a minimum depth of concrete over the length of the opening of 8 inches (203 mm) in flat and waffle-grid ICF walls and 12 inches (305 mm) in screen-grid ICF walls. Wall openings in waffle- and screen-grid ICF walls shall be located such that concrete vertical cores, with a minimum cross section equal to T x $0.5W$, occur along each side of the opening. Refer to Section 2.0 for the definitions of variables, T and W. Reinforcement around openings shall be provided in accordance with Table 5.4 and Figure 5.2 in addition to the minimum vertical wall reinforcement required in Sections 3.0 and 4.0.

Exception: Continuous horizontal wall reinforcement placed within 12 inches of the top of the wall story as required in Sections 3.0 and 4.0 is permitted to be used in lieu of top or bottom lintel reinforcement provided that the continuous horizontal wall reinforcement meets the location requirements specified in Figures 5.3 and 5.4 and the size requirements specified in Tables 5.5 through 5.10.

All opening reinforcement placed horizontally above or below an opening shall extend a minimum of 24 inches (610 mm) beyond the limits of the opening. Where 24 inches (610 mm) cannot be obtained beyond the limit of the opening, the bar may be bent 90 degrees in order to obtain adequate embedment. In conditions with wind speeds greater than 90 mph (145 km/hr), all vertical opening reinforcement in the top-most ICF story shall be terminated with a minimum 90 degree bend. The bend shall extend a minimum of 24 inches (610 mm) into the lintel parallel to the top and bottom lintel reinforcement and lie within 4 inches (102 mm) of the top surface of the ICF wall near roof level.

5.3 Lintels

5.3.1 Load-Bearing ICF Wall Lintels

Lintels shall be provided in load-bearing walls over all openings greater than or equal to 4 feet (1.2 m) in width. Lintel depth, D, is permitted to include the available height of any ICF wall located directly above the lintel, provided that the increased lintel depth spans the entire length of the lintel. A minimum of one No. 3 stirrup shall be installed for all flat ICF lintels with spans greater than 4 feet (1.2 m) at a maximum spacing of *d/2*, where *d* is defined in Figure 5.3. A minimum of two No. 3 stirrups shall be installed in each vertical concrete core for all waffle-grid lintels with spans greater than 4 feet (1.2 m).

Lintels for flat ICF walls shall be constructed in accordance with Figure 5.3 and Tables 5.5 and 5.6. Lintels for waffle-grid ICF walls shall be constructed in accordance with Figure 5.4 and Tables 5.7 and 5.8. Lintels for flat and waffle-grid ICF walls spanning between 12 feet (3.7 m) and 16 feet (4.9 m) shall be constructed in accordance with Table 5.9. Lintels for screen-grid ICF walls shall be constructed in accordance with the requirements for flat ICF walls. Lintels spanning greater than or equal to 4 feet (1.2 m) using screen-grid ICFs shall not be permitted in load-bearing ICF walls.

5.3.2 ICF Gable End Wall Lintels (Non Load-Bearing)

Lintels shall be provided in gable end walls over all openings greater than or equal to 4 feet (1.2 m) in width. Lintel depth, D, is permitted to include the available height of the ICF wall located directly above the lintel, provided that the increased lintel depth spans the entire length of the lintel. A minimum of one No. 3 stirrup shall be installed for all flat ICF lintels with spans greater than 4 feet (1.2 m) at a maximum spacing of *d/2*, where *d* is defined in Figure 5.3. A minimum of two No. 3 stirrups shall be installed in each vertical concrete core for all waffle-grid lintels with spans greater than 4 feet (1.2 m).

Lintels for flat ICF walls shall be constructed in accordance with Figure 5.3 and Table 5.10. Lintels for waffle-grid ICF walls shall be constructed in accordance with Figure 5.4 and Table 5.10. Lintels for screen-grid ICF wall openings shall be constructed in accordance with the requirements for flat ICF walls. Lintels spanning greater than or equal to 4 feet (1.2 m) using screen-grid ICFs shall not be permitted.

TABLE 5.1
ADJUSTMENT FACTORS FOR USE WITH TABLE 5.2[1]

Building Aspect Ratio[2]	Adjustment Factor	
	Parallel to the Shorter Building Dimension or Endwall, W	Parallel to the Longer Building Dimension or Sidewall, L
2.0	1.0	0.25
1.8	0.9	0.30
1.6	0.8	0.35
1.4	0.7	0.40
1.2	0.6	0.45
1.0	0.5	0.50

1 Linear interpolation between building aspect ratios shall be permitted.
2 Refer to Figure 5.1 for the definition of the building aspect ratio.

TABLE 5.2
MINIMUM PERCENTAGE OF SOLID WALL LENGTH ALONG EXTERIOR WALL LINES[1]

Exterior Wall Line Category	Site Wind and Seismic Conditions[2,3]	
	All Wind Speeds and Seismic Zones 0 and 1	All Wind Speeds and Seismic Zone 2
Parallel to the Shorter Building Dimension or Endwall, W	The minimum percentage of solid wall length shall be the maximum of the following: • The base percentage of solid wall length based on wind speed in Table 5.3 multiplied by the appropriate factor from Table 5.1. • 15 percent for walls supporting a light-frame roof or 20 percent for walls supporting an ICF or light-frame second story and light-frame roof.	The minimum percentage of solid wall length shall be the maximum of the following: • The base percentage of solid wall length based on wind speed in Table 5.3 multiplied by the appropriate factor from Table 5.1. • The base percentage of solid wall length for Seismic Zone 2 in Table 5.3 without adjustment. • 15 percent for walls supporting a light-frame roof or 20 percent for walls supporting an ICF or light-frame second story and light-frame roof .
Parallel to the Longer Building Dimension or Sidewall, L	The minimum percentage of solid wall length shall be the maximum of the following: • The base percentage of solid wall length based on wind speed in Table 5.3 multiplied by the appropriate factor from Table 5.1. • 15 percent for walls supporting a light-frame roof or 20 percent for walls supporting an ICF or light-frame second story and light-frame roof.	The minimum percentage of solid wall length shall be the maximum of the following: • The base percentage of solid wall length based on wind speed in Table 5.3 multiplied by the appropriate factor from Table 5.1. • The base percentage of solid wall length for Seismic Zone 2 in Table 5.3 divided by the building aspect ratio. • 15 percent for walls supporting a light-frame roof or 20 percent for walls supporting an ICF or light-frame second story and light-frame roof.

1 Refer to Figure 5.1 for the definition of L, W, and the building aspect ratio.
2 The percent of solid wall in Seismic Zones 3 and 4, regardless of wind speed, shall be determined by an approved design.
3 In no case shall the spacing of wall segments at least 24 inches (610 mm) in length exceed 18 feet (5.5 m) on center.

TABLE 5.3
BASE PERCENTAGE OF SOLID WALL LENGTH ALONG EXTERIOR WALL LINES[1,2,3]

ICF Wall Type and Minimum Wall Thickness (inches)[4]	Max. Roof Slope	Base Solid Wall Length (percent)									
		Wall Supporting Light-Frame Roof Only					Wall Supporting ICF or Light-Frame Second Story and Light-Frame Roof				
		Maximum Wind Speed (mph)									
		70	80	90	100	110	70	80	90	100	110
Flat 3.5	3:12	15	15	15	15	20	30	35	40	50	55
	6:12	15	15	20	20	25	30	40	50	55	60
	9:12	20	25	30	40	45	45	60	70	85	95
	12:12	25	35	40	50	60	50	65	80	95	100
Flat 5.5	3:12	15	15	15	15	15	20	25	30	40	40
	6:12	15	15	15	15	20	20	30	35	40	45
	9:12	15	15	20	25	30	35	45	50	60	70
	12:12	20	20	25	35	40	35	50	55	70	75
Flat 7.5	3:12	15	15	15	15	15	20	20	25	30	30
	6:12	15	15	15	15	15	20	20	25	30	35
	9:12	15	15	15	20	25	25	30	40	45	50
	12:12	15	20	20	25	30	30	35	40	50	55
Waffle-Grid 6	3:12	15	15	15	15	20	25	30	35	45	50
	6:12	15	15	20	20	25	25	35	45	50	55
	9:12	20	20	25	35	40	40	55	60	75	85
	12:12	25	30	35	45	50	45	60	70	85	90
Waffle-Grid 8	3:12	15	15	15	15	15	20	25	30	35	35
	6:12	15	15	15	15	20	20	25	30	35	40
	9:12	15	15	20	25	30	30	40	45	55	60
	12:12	20	20	25	30	35	35	40	50	60	65
Screen-Grid 6	3:12	15	15	20	20	25	30	40	45	55	60
	6:12	15	20	25	30	35	30	40	50	60	70
	9:12	25	30	45	50	65	50	65	80	90	100
	12:12	35	40	55	65	80	55	70	85	100	100
Seismic Zone 2											
Flat, 3.5	N/A	20					35				
Flat, 5.5	N/A	15					30				
Flat, 7.5	N/A	15					25				
Waffle-Grid, 6	N/A	20					35				
Waffle-Grid, 8	N/A	20					30				
Screen-Grid, 6	N/A	25					45				

For SI: 1 inch = 25.4 mm; 1 mph = 1.6093 km/hr

[1] Linear interpolation between roof slopes shall be permitted.
[2] Base percentages are applicable for maximum 10-foot (3.0-m) wall story heights.
[3] N/A indicates not applicable.
[4] Actual thickness is shown for flat walls while nominal thickness is given for waffle- and screen-grid walls. Refer to Section 2.0 for actual waffle- and screen-grid thickness and dimensions.

TABLE 5.4
MINIMUM WALL OPENING REINFORCEMENT REQUIREMENTS IN ICF WALLS

Wall Type and Opening Width, L feet (m)	Minimum Horizontal Opening Reinforcement	Minimum Vertical Opening Reinforcement
Flat, Waffle-, and Screen-Grid: L < 2 (0.61)	None Required	None Required
Screen-Grid: 2 (0.61) ≤ L ≤ 4 (1.2)	One No. 4 bar a minimum of 1.5 inches (38 mm) and a maximum of 2.5 inches (64 mm) from the top of the opening. One No. 4 bar within 12 inches (305 mm) of the bottom of the opening. Each No. 4 bar shall extend a minimum of 24 inches (610 mm) beyond the limits of the opening.	None Required
Flat and Waffle-Grid: 2 (0.61) ≤ L < 4 (1.2)	One No. 4 bar within 12 inches (305 mm) of the top of the opening. One No. 4 bar within 12 inches (305 mm) from the bottom of the opening. Each No. 4 bar shall extend a minimum of 24 inches (610 mm) beyond the limits of the opening.	None Required
Flat, Waffle-, and Screen-Grid: L ≥ 4 (1.2)	Provide lintels in accordance with Section 5.0. Top and bottom lintel reinforcement shall extend a minimum of 24 inches (610 mm) beyond the limits of the opening.	In conditions with wind speeds less than or equal to 90 mph (145 km/hr), provide one No. 4 bar for the full height of the wall story within 12 inches (305 mm) of each side of the opening. In conditions with wind speeds greater than 90 mph (145 km/hr), provide two No. 4 bars for the full height of the wall story within 12 inches (305 mm) of each side of the opening.

TABLE 5.5
MAXIMUM ALLOWABLE CLEAR SPANS FOR FLAT ICF LINTELS IN LOAD-BEARING WALLS[1,2,3]
NO. 4 BOTTOM BAR SIZE

Minimum Lintel Thickness, T (inches)	Minimum Lintel Depth, D (inches)	Maximum Clear Span					
		Supporting Light-Frame Roof Only		Supporting Light-Frame Second Story and Roof		Supporting ICF Second Story and Light-Frame Roof [4]	
		Maximum Ground Snow Load (psf)					
		30	70	30	70	30	70
3.5	8	4'-9"	4'-2"	3'-10"	3'-4"	3'-5"	3'-1"
	12	6'-8"	5'-5"	5'-0"	4'-5"	4'-6"	4'-0"
	16	7'-11"	6'-5"	6'-0"	5'-3"	5'-4"	4'-10"
	20	8'-11"	7'-4"	6'-9"	6'-0"	6'-1"	5'-6"
	24	9'-10"	8'-1"	7'-6"	6'-7"	6'-9"	6'-1"
5.5	8	5'-2"	4'-2"	3'-10"	3'-5"	3'-5"	3'-1"
	12	6'-8"	5'-5"	5'-0"	4'-5"	4'-6"	4'-1"
	16	7'-10"	6'-5"	6'-0"	5'-3"	5'-4"	4'-10"
	20	8'-10"	7'-3"	6'-9"	6'-0"	6'-1"	5'-6"
	24	9'-8"	8'-0"	7'-5"	6'-7"	6'-8"	6'-0"
7.5	8	5'-2"	4'-2"	3'-11"	3'-5"	3'-6"	3'-2"
	12	6'-7"	5'-5"	5'-0"	4'-5"	4'-6"	4'-1"
	16	7'-9"	6'-5"	5'-11"	5'-3"	5'-4"	4'-10"
	20	8'-8"	7'-2"	6'-8"	5'-11"	6'-0"	5'-5"
	24	9'-6"	7'-11"	7'-4"	6'-6"	6'-7"	6'-0"
9.5	8	5'-2"	4'-2"	3'-11"	3'-5"	3'-6"	3'-2"
	12	6'-7"	5'-5"	5'-0"	4'-5"	4'-6"	4'-1"
	16	7'-8"	6'-4"	5'-11"	5'-3"	5'-4"	4'-10"

For SI: 1 foot = 0.3048 m; 1 inch = 25.4 mm; 1 psf = 47.8804 Pa

1 Deflection criterion is *L/240*, where *L* is the clear span of the lintel in inches.
2 Linear interpolation is permitted between ground snow loads and between lintel depths.
3 No. 3 stirrups are required at maximum *d/2* spacing for spans greater than 4 feet (1.2 m).
4 ICF wall dead load is maximum 69 psf (3.3 kPa).

TABLE 5.6
MAXIMUM ALLOWABLE CLEAR SPANS FOR FLAT ICF LINTELS IN LOAD-BEARING WALLS[1,2,3]
NO. 5 BOTTOM BAR SIZE

Minimum Lintel Thickness, T (inches)	Minimum Lintel Depth, D (inches)	Maximum Clear Span					
		Supporting Light-Frame Roof Only		Supporting Light-Frame Second Story and Roof		Supporting ICF Second Story and Light-Frame Roof[4]	
		Maximum Ground Snow Load (psf)					
		30	70	30	70	30	70
3.5	8	4'-9"	4'-2"	3'-11"	3'-7"	3'-7"	3'-5"
	12	7'-2"	6'-3"	5'-11"	5'-5"	5'-5"	5'-0"
	16	9'-6"	8'-0"	7'-4"	6'-6"	6'-7"	5'-11"
	20	11'-1"	9'-1"	8'-4"	7'-5"	7'-6"	6'-9"
	24	12'-2"	10'-0"	9'-3"	8'-2"	8'-4"	7'-6"
5.5	8	5'-6"	4'-10"	4'-7"	4'-2"	4'-2"	3'-10"
	12	8'-3"	6'-9"	6'-3"	5'-6"	5'-7"	5'-0"
	16	9'-9"	8'-0"	7'-5"	6'-6"	6'-7"	6'-0"
	20	10'-11"	9'-0"	8'-4"	7'-5"	7'-6"	6'-9"
	24	12'-0"	9'-11"	9'-3"	8'-2"	8'-3"	7'-6"
7.5	8	6'-1"	5'-2"	4'-9"	4'-3"	4'-3"	3'-10"
	12	8'-2"	6'-9"	6'-3"	5'-6"	5'-7"	5'-0"
	16	19'-7"	7'-11"	7'-4"	6'-6"	6'-7"	6'-0"
	20	10'-10"	8'-11"	8'-4"	7'-4"	7'-6"	6'-9"
	24	11'-10"	9'-10"	9'-2"	8'-1"	8'-3"	7'-5"
9.5	8	6'-4"	5'-2"	4'-10"	4'-3"	4'-4"	3'-11"
	12	8'-2"	6'-8"	6'-2"	5'-6"	5'-7"	5'-0"
	16	9'-6"	7'-11"	7'-4"	6'-6"	6'-7"	5'-11"
	20	10'-8"	8'-10"	8'-3"	7'-4"	7'-5"	6'-9"
	24	11'-7"	9'-9"	9'-0"	8'-1"	8'-2"	7'-5"

For SI: 1 foot = 0.3048 m; 1 inch = 25.4 mm; 1 psf = 47.8804 Pa

[1] Deflection criterion is *L/240*, where *L* is the clear span of the lintel in inches.
[2] Linear interpolation is permitted between ground snow loads and between lintel depths.
[3] No. 3 stirrups are required at maximum *d/2* spacing for spans greater than 4 feet (1.2 m).
[4] ICF wall dead load is maximum 69 psf (3.3 kPa).

TABLE 5.7
MAXIMUM ALLOWABLE CLEAR SPANS FOR WAFFLE-GRID ICF LINTELS IN LOAD-BEARING WALLS[1,2,3]
NO. 4 BOTTOM BAR SIZE

Minimum Lintel Thickness, T[4] (inches)	Minimum Lintel Depth, D (inches)	Maximum Clear Span					
		Supporting Light-Frame Roof Only		Supporting Light-Frame Second Story and Roof		Supporting ICF Second Story and Light-Frame Roof[5]	
		Maximum Ground Snow Load (psf)					
		30	70	30	70	30	70
6	8	5'-2"	4'-2"	3'-10"	3'-5"	3'-6"	3'-2"
	12	6'-8"	5'-5"	5'-0"	4'-5"	4'-7"	4'-2"
	16	7'-11"	6'-6"	6'-0"	5'-3"	5'-6"	4'-11"
	20	8'-11"	7'-4"	6'-9"	6'-0"	6'-3"	5'-7"
	24	9'-10"	8'-1"	7'-6"	6'-7"	6'-10"	6'-2"
8	8	5'-2"	4'-3"	3'-11"	3'-5"	3'-7"	3'-2"
	12	6'-8"	5'-5"	5'-1"	4'-5"	4'-8"	4'-2"
	16	7'-10"	6'-5"	6'-0"	5'-3"	5'-6"	4'-11"
	20	8'-10"	7'-3"	6'-9"	6'-0"	6'-2"	5'-7"
	24	9'-8"	8'-0"	7'-5"	6'-7"	6'-10"	6'-2"

For SI: 1 foot = 0.3048 m; 1 inch = 25.4 mm; 1 psf = 47.8804 Pa

1 Deflection criterion is *L/240*, where *L* is the clear span of the lintel in inches.
2 Linear interpolation is permitted between ground snow loads and between lintel depths.
3 Two No. 3 stirrups are required in each vertical concrete core for spans greater than 4 feet (1.2 m).
4 Lintel thickness corresponds to the nominal waffle-grid ICF wall thickness with a minimum web thickness of 2 inches (51 mm). For actual wall thickness, refer to Section 2.0.
5 ICF wall dead load is maximum 55 psf (2.6 kPa).

TABLE 5.8
MAXIMUM ALLOWABLE CLEAR SPANS FOR WAFFLE-GRID ICF LINTELS IN LOAD-BEARING WALLS[1,2,3]
NO. 5 BOTTOM BAR SIZE

Minimum Lintel Thickness, T[4] (inches)	Minimum Lintel Depth, D (inches)	Maximum Clear Span					
		Supporting Light-Frame Roof Only		Supporting Light-Frame Second Story and Roof		Supporting ICF Second Story and Light-Frame Roof[5]	
		Maximum Ground Snow Load (psf)					
		30	70	30	70	30	70
6	8	5'-4"	4'-8"	4'-5"	4'-1"	4'-5"	3'-10"
	12	8'-0"	6'-9"	6'-3"	5'-6"	6'-3"	5'-1"
	16	9'-9"	8'-0"	7'-5"	6'-6"	7'-5"	6'-1"
	20	11'-0"	9'-1"	8'-5"	7'-5"	8'-5"	6'-11"
	24	12'-2"	10'-0"	9'-3"	8'-2"	9'-3"	7'-8"
8	8	6'-0"	5'-2"	4'-9"	4'-3"	4'-9"	3'-11"
	12	8'-3"	6'-9"	6'-3"	5'-6"	6'-3"	5'-2"
	16	9'-9"	8'-0"	7'-5"	6'-6"	7'-5"	6'-1"
	20	10'-11"	9'-0"	8'-4"	7'-5"	8'-4"	6'-11"
	24	12'-0"	9'-11"	9'-2"	8'-2"	9'-2"	7'-8"

FOR SI: 1 inch = 25.4 mm; 1 psf = 0.0479 kN/m^2; 1 ft = 0.3 m

1 Deflection criterion is *L/240*, where *L* is the clear span of the lintel in inches.
2 Interpolation is permitted between ground snow loads and between lintel depths.
3 Two No. 3 stirrups are required in each vertical concrete core for spans greater than 4 feet (1.2 m).
4 Lintel thickness corresponds to the nominal waffle-grid ICF wall thickness with a minimum web thickness of 2 inches (51 mm). For actual wall thickness, refer to Section 2.0.
5 ICF wall dead load is maximum 55 psf (2.6 kPa).

TABLE 5.9
MINIMUM BOTTOM BAR ICF LINTEL REINFORCEMENT FOR LARGE CLEAR SPANS IN LOAD-BEARING WALLS[1,2,3,4]

Minimum Lintel Thickness, T[5] (inches)	Minimum Lintel Depth, D (inches)	Minimum Bottom Lintel Reinforcement					
		Supporting Light-Frame Roof Only		Supporting Light-Frame Second Story and Roof		Supporting ICF Second Story and Light-Frame Roof[6]	
		Maximum Ground Snow Load (psf)					
		30	70	30	70	30	70
Flat ICF Lintel, 12 Feet Maximum Clear Span							
3.5	24	1-#5	1-#7	D/R	D/R	D/R	D/R
5.5	20	1-#6; 2-#4	1-#7; 2-#5	D/R	D/R	D/R	D/R
	24	1-#5	1-#7; 2-#5	1-#7; 2-#5	1-#8; 2-#6	1-#8; 2-#6	D/R
7.5	16	1-#7; 2-#5	D/R	D/R	D/R	D/R	D/R
	20	1-#6; 2-#4	1-#7; 2-#5	1-#8; 2-#6	D/R	D/R	D/R
	24	1-#6; 2-#4	1-#7; 2-#5	1-#7; 2-#5	1-#8; 2-#6	1-#8; 2-#6	1-#8; 2-#6
9.5	16	1-#7; 2-#5	D/R	D/R	D/R	D/R	D/R
	20	1-#6; 2-#4	1-#7; 2-#5	1-#8; 2-#6	1-#8; 2-#6	1-#8; 2-#6	1-#9; 2-#6
	24	1-#6; 2-#4	1-#7; 2-#5	1-#7; 2-#5	1-#8; 2-#6	1-#8; 2-#6	1-#9; 2-#6
Flat ICF Lintel, 16 Feet Maximum Clear Span							
5.5	24	1-#7; 2-#5	D/R	D/R	D/R	D/R	D/R
7.5	24	1-#7; 2-#5	D/R	D/R	D/R	D/R	D/R
9.5	24	1-#7; 2-#5	1-#9; 2-#6	1-#9; 2-#6	D/R	D/R	D/R
Waffle-Grid ICF Lintel, 12 Feet Maximum Clear Span							
6	20	1-#6; 2-#4	D/R	D/R	D/R	D/R	D/R
	24	1-#5	1-#7; 2-#5	1-#7; 2-#5	1-#8; 2-#6	1-#8; 2-#6	D/R
8	16	1-#7; 2-#5	D/R	D/R	D/R	D/R	D/R
	20	1-#6; 2-#4	1-#7; 2-#5	1-#8; 2-#6	D/R	D/R	D/R
	24	1-#5	1-#7; 2-#5	1-#7; 2-#5	1-#8; 2-#6	1-#8; 2-#6	1-#8; 2-#6

FOR SI: 1 inch = 25.4 mm; 1 psf = 0.0479 kN/m^2; 1 ft = 0.3 m

[1] D/R indicates design is required.
[2] Deflection criterion is *L/240*, where *L* is the clear span of the lintel in inches.
[3] Interpolation is permitted between ground snow loads and between lintel depths.
[4] No. 3 stirrups are required for spans greater than 4 feet (1.2 m). Flat ICF lintels require one No. 3 stirrup at maximum *d/2* spacing and waffle-grid ICF lintels require two No. 3 stirrups in each vertical concrete core.
[5] Actual thickness is shown for flat lintels while nominal thickness is given for waffle-grid lintels. Lintel thickness corresponds to the nominal waffle-grid ICF wall thickness with a minimum web thickness of 2 inches (51 mm). Refer to Section 2.0 for actual wall thickness.
[6] ICF wall dead load is maximum 55 psf (2.6 kPa).

TABLE 5.10
MAXIMUM ALLOWABLE CLEAR SPANS FOR ICF LINTELS IN GABLE END (NON-LOAD-BEARING) WALLS[1,2,3]
NO. 4 BOTTOM BAR SIZE

Minimum Lintel Depth, D (inches)	Maximum Clear Span	
	Supporting Light-Frame Gable End Wall (feet)	Supporting ICF Second Story Gable End Wall[4] (feet)
8	12	6
12	16	8
16	16	10
20	16	12
24	16	16

For SI: 1 foot = 0.3048 m; 1 inch = 25.4 mm; 1 psf = 47.8804 Pa

[1] Deflection criterion is *L/240*, where *L* is the clear span of the lintel in inches.
[2] Linear interpolation is permitted between lintel depths.
[3] No. 3 stirrups are required for spans greater than 4 feet (1.2 m). Flat ICF lintels require one No. 3 stirrup at maximum *d/2* spacing and waffle-grid ICF lintels require two No. 3 stirrups in each vertical concrete core.
[4] ICF wall dead load is maximum 69 psf (3.3 kPa).

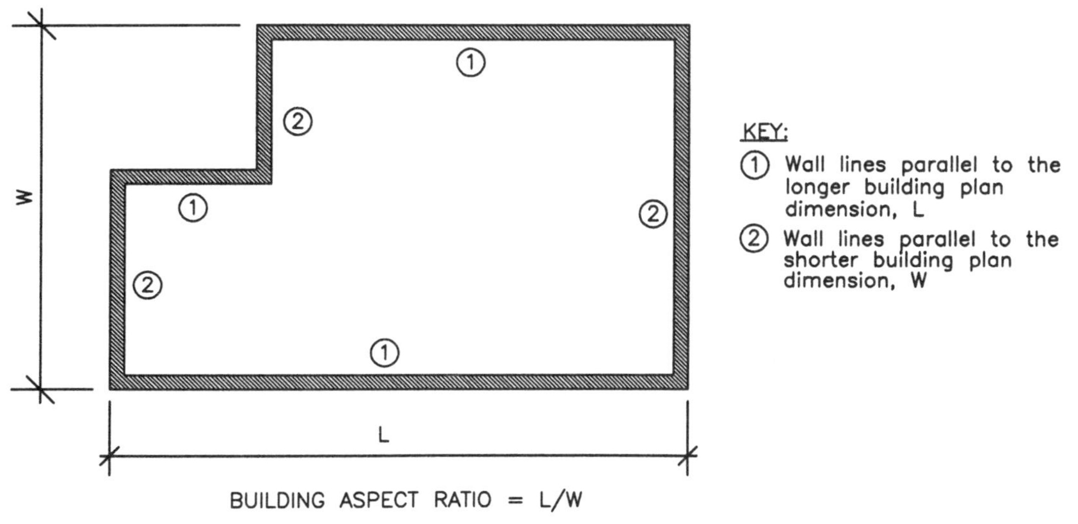

Figure 5.1 Building Aspect Ratio

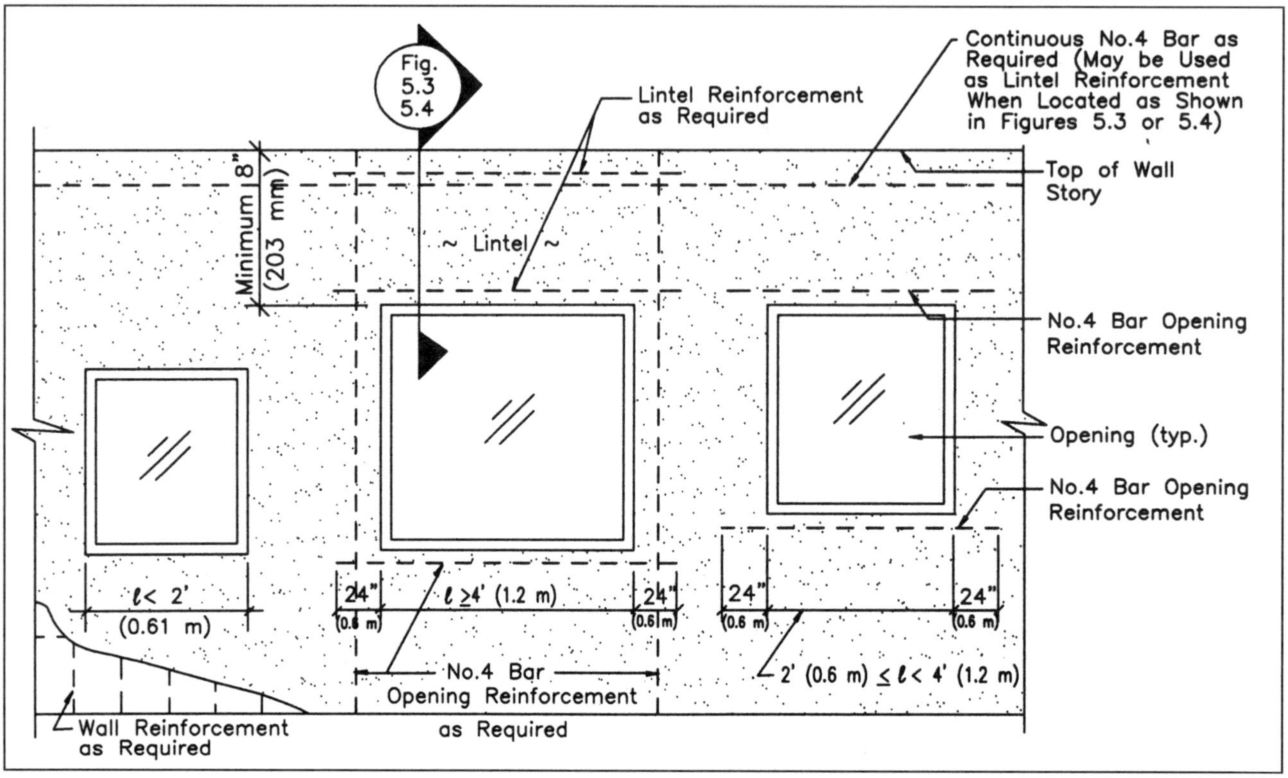

Figure 5.2 Reinforcement of Openings

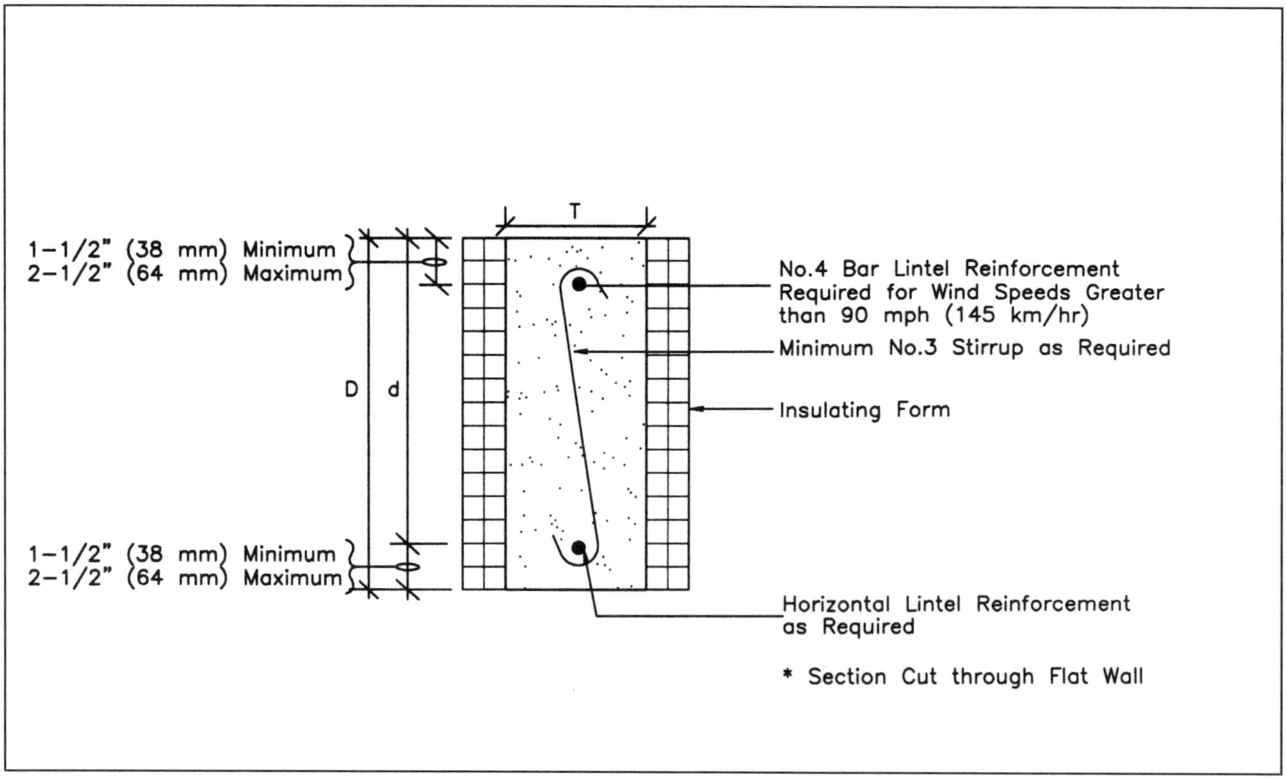

Figure 5.3 Flat ICF Lintel Construction

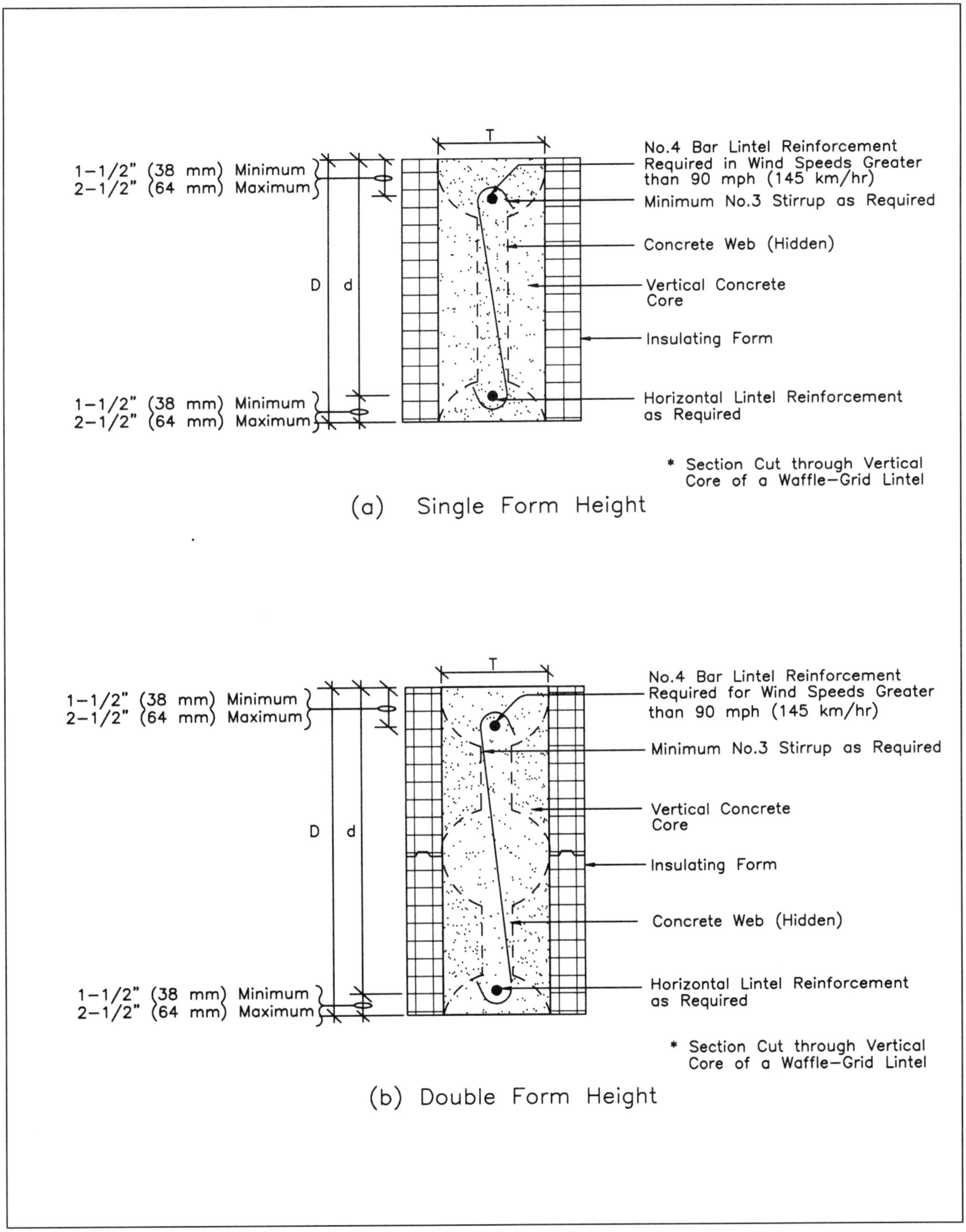

Figure 5.4 Waffle-Grid ICF Lintel Construction

6.0 ICF CONNECTION REQUIREMENTS

All ICF walls shall be connected to footings, floors, and roofs in accordance with this section. Requirements for installation of brick veneer on exterior ICF walls and other construction details not covered in this section shall comply with the manufacturer's recommendations and sound practice.

6.1 ICF Foundation Wall-to-Footing Connection

No vertical reinforcement (i.e., dowel) across the joint between the foundation wall and the footing is required when one of the following exists:

- The unbalanced backfill height does not exceed 4 feet (1.2 m).
- The interior floor slab is installed in accordance with Figure 3.3 before backfilling.
- Temporary bracing at the bottom of the foundation wall is erected before backfilling and remains in place during construction until an interior floor slab is installed in accordance with Figure 3.3 or the wall is backfilled on both sides (i.e., stem wall).

For foundation walls that do not meet one of the above requirements, vertical reinforcement (i.e., dowel) shall be installed across the joint between the foundation wall and the footing in accordance with Figure 6.1.

Exception: The foundation wall's vertical wall reinforcement, at intervals of 4 feet (1.2 m) on center, shall extend 8 inches (203 mm) into the footing in lieu of using a dowel as shown in Figure 6.1.

6.2 ICF Wall-to-Floor Connection

6.2.1 Floor on ICF Wall Connection (Top-Bearing Connection)

Floors bearing on ICF walls shall be constructed in accordance with Figure 6.2 or 6.3. The wood sill plate or floor system shall be anchored to the ICF wall with 1/2-inch- (13-mm-) diameter bolts placed at a maximum spacing of 6 feet (1.8 m) on center and not more than 12 inches (305 mm) from joints in the sill plate. Anchor bolts shall extend a minimum of 7 inches (178 mm) into the concrete. Light-frame construction shall be in accordance with the applicable building code. In conditions where wind speeds are 90 mph (145 km/hr) or greater, the 1/2-inch- (13-mm-) diameter anchor bolts shall be placed at a maximum spacing of 4 feet (1.2 m) on center.

6.2.2 Floor Ledger-ICF Wall Connection (Side-Bearing Connection)

Wood ledger boards shall be anchored to flat ICF walls having a minimum thickness of 5.5 inches (140 mm) thickness and to waffle- or screen-grid ICF walls having a minimum nominal thickness of 6 inches (152 mm) in accordance with Figure 6.4 or 6.5 and Table 6.1. Wood ledger boards shall be anchored to flat ICF walls having a minimum thickness of 3.5 inches (89 mm) in accordance with Figure 6.6 or 6.7 and Table 6.1.

6.3 ICF Wall-to-Roof Connection

Wood sill plates attaching roof framing to ICF walls shall be anchored to the ICF wall in accordance with Table 6.2 and Figure 6.8. Anchor bolts shall be located in the center of the flat ICF wall thickness or the center of the vertical core thickness of the waffle-grid and screen-grid ICF wall system and shall have a minimum embedment of 4 inches (102 mm). Roof framing attachment to wood sill plates shall be in accordance with the local building code.

In conditions where wind speeds are 90 mph (145 km/hr) or greater, an approved uplift connector (i.e., strap or bracket) shall be used to attach roof assemblies to wood sill plates in accordance with Table 6.3 and Figure 6.9 or directly to ICF walls in accordance with Table 6.3 and Figure 6.10. Embedment of strap connectors shall be in accordance with the strap connector manufacturer's recommendations.

TABLE 6.1
FLOOR LEDGER-ICF WALL CONNECTION (SIDE-BEARING CONNECTION) REQUIREMENTS[1,2,3]

Maximum Floor Clear Span[4] (feet)	Maximum Anchor Bolt Spacing[5] (inches)			
	Staggered 1/2-Inch-Diameter Anchor Bolts	Staggered 5/8-Inch-Diameter Anchor Bolts	Two 1/2-Inch-Diameter Anchor Bolts[6]	Two 5/8-Inch-Diameter Anchor Bolts[6]
8	18	20	36	40
10	16	18	32	36
12	14	18	28	36
14	12	16	24	32
16	10	14	20	28
18	9	13	18	26
20	8	11	16	22
22	7	10	14	20
24	7	9	14	18
26	6	9	12	18
28	6	8	12	16
30	5	8	10	16
32	5	7	10	14

For SI: 1 foot = 0.3048 m; 1 inch = 25.4 mm

1 Minimum ledger board nominal depth shall be 8 inches (203 mm). The thickness of the ledger board shall be a minimum of 2 inches (51 mm). Thickness of ledger board is in nominal lumber dimensions. Ledger board shall be minimum No. 2 Grade.

2 Minimum edge distance shall be 2 inches (51 mm) for 1/2-inch- (13-mm-) diameter anchor bolts and 2.5 inches (64 mm) for 5/8-inch- (16-mm-) diameter anchor bolts.

3 Interpolation is permitted between floor spans.

4 Floor span corresponds to the clear span of the floor structure (i.e., joists or trusses) spanning between load-bearing walls or beams.

5 Anchor bolts shall extend through the ledger to the center of the flat ICF wall thickness or the center of the horizontal or vertical core thickness of the waffle-grid or screen-grid ICF wall system.

6 Minimum vertical distance between bolts shall be 1.5 inches (38 mm) for 1/2-inch- (13-mm-) diameter anchor bolts and 2 inches (51 mm) for 5/8-inch- (16-mm-) diameter anchor bolts.

TABLE 6.2
TOP SILL PLATE-ICF WALL CONNECTION REQUIREMENTS

Maximum Wind Speed (mph)	Maximum Anchor Bolt Spacing [1]	
	1/2-Inch-Diameter Anchor Bolt	5/8-Inch-Diameter Anchor Bolt
70	6'-0"	8'-0"
80	6'-0"	8'-0"
90	6'-0"	8'-0"
100	4'-0"	6'-0"
110	4'-0"	4'-0"

For SI: 1 foot = 0.3048 m; 1 inch = 25.4 mm; 1 mph = 16.09344 km/hr

[1] Minimum anchor bolt embedment length shall be 4 inches (102 mm).

TABLE 6.3
ROOF STRAP UPLIFT REQUIREMENTS[1,2]

Maximum Roof Clear Span (feet)	Maximum Number of Stories Above Grade	Minimum Strap Uplift Load (lb) Maximum Wind Speed (mph)			
		80[3]	90	100	110
24	One	240	400	570	760
	Two	320	490	680	900
28	One	270	450	640	860
	Two	350	550	770	1010
32	One	310	500	720	960
	Two	390	600	850	1120
36	One	340	560	800	1080
	Two	430	670	940	1240
40	One	380	620	890	1190
	Two	470	740	1030	1360

For SI: 1 foot = 0.3048 m; 1 inch = 25.4 mm; 1 mph = 16.09344 km/hr

[1] Maximum roof overhang shall be 2 feet (1.2 m).
[2] Maximum roof truss or rafter spacing shall be 2 feet (1.2 m) on center. For 16-inch (406-mm) and 12-inch (305-mm) truss or rafter spacing, the required loads shall be multiplied by 0.67 and 0.5, respectively.
[3] Strap is not required when roof framing is connected to a wood sill plate in accordance with the local building code and the wood sill plate is anchored to the wall in accordance with Table 6.2.

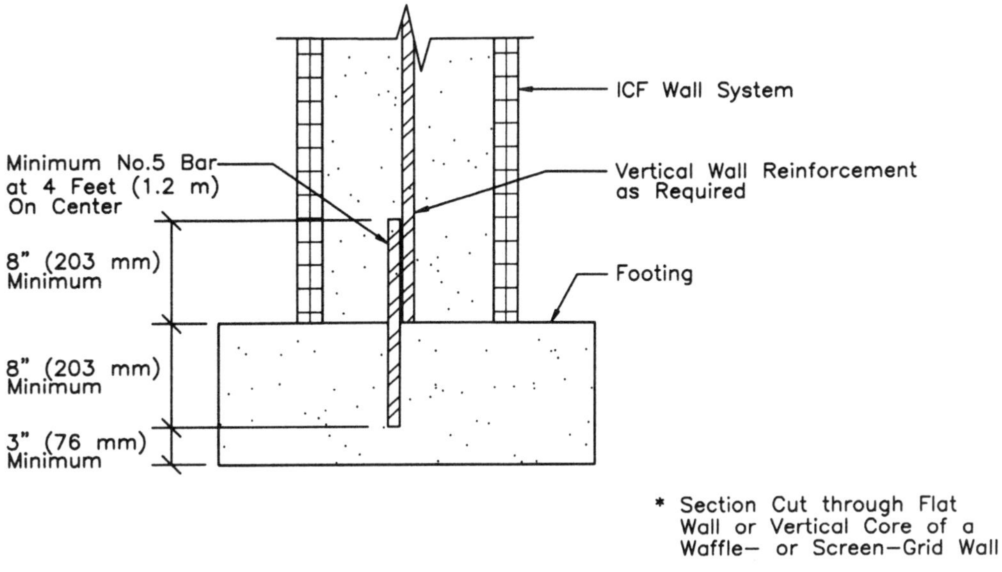

Figure 6.1 ICF Foundation Wall-to-Footing Connection

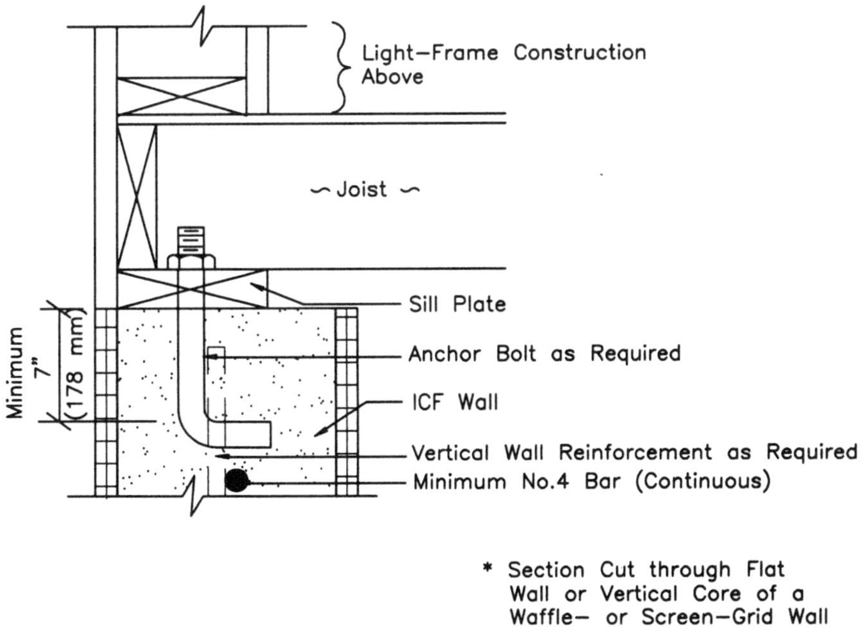

Figure 6.2 Floor on ICF Wall Connection (Top-Bearing Connection)

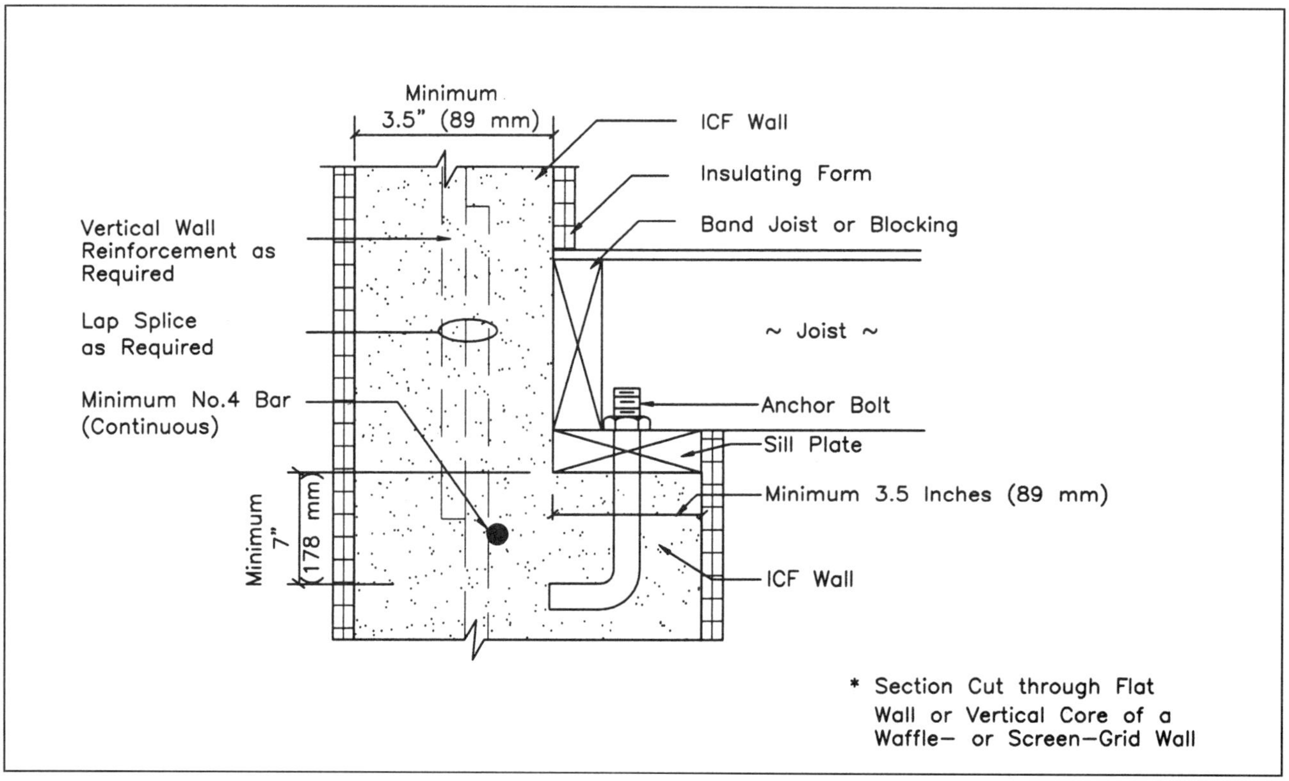

Figure 6.3 Floor on ICF Wall Connection (Top-Bearing Connection)

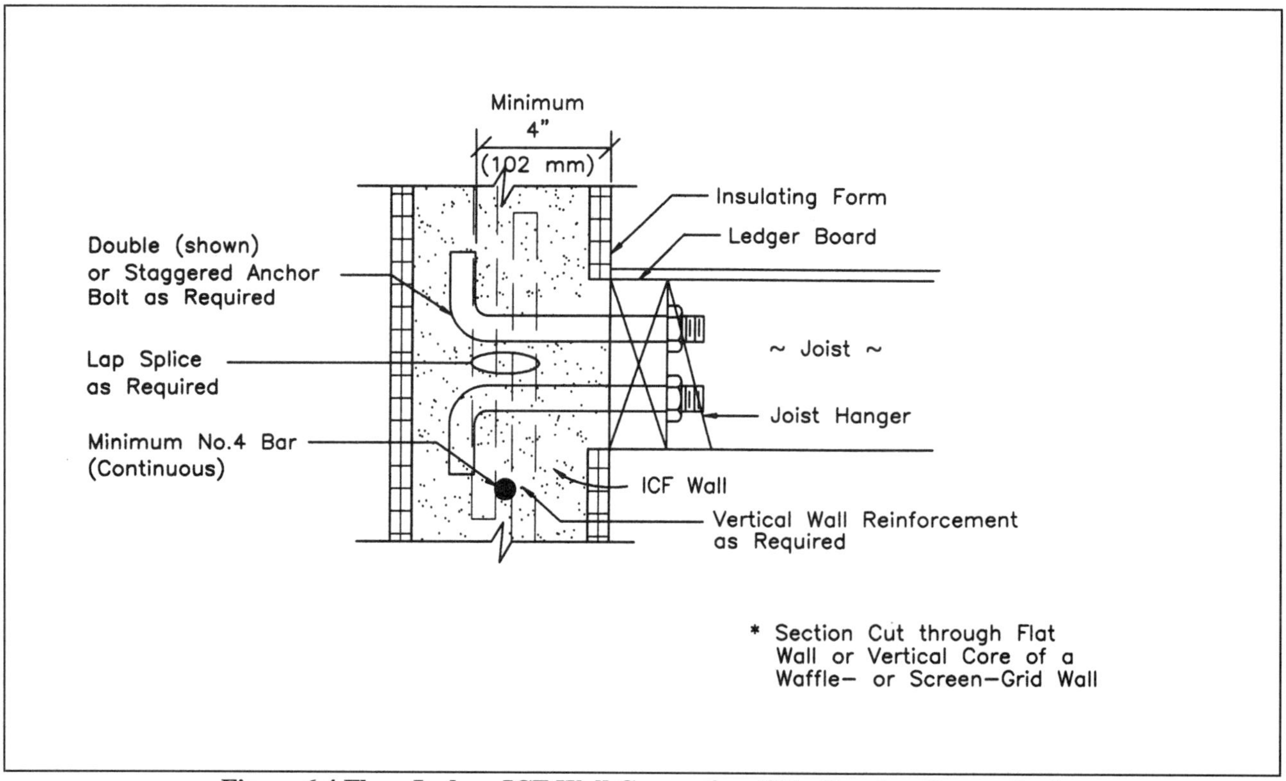

Figure 6.4 Floor Ledger-ICF Wall Connection (Side-Bearing Connection)

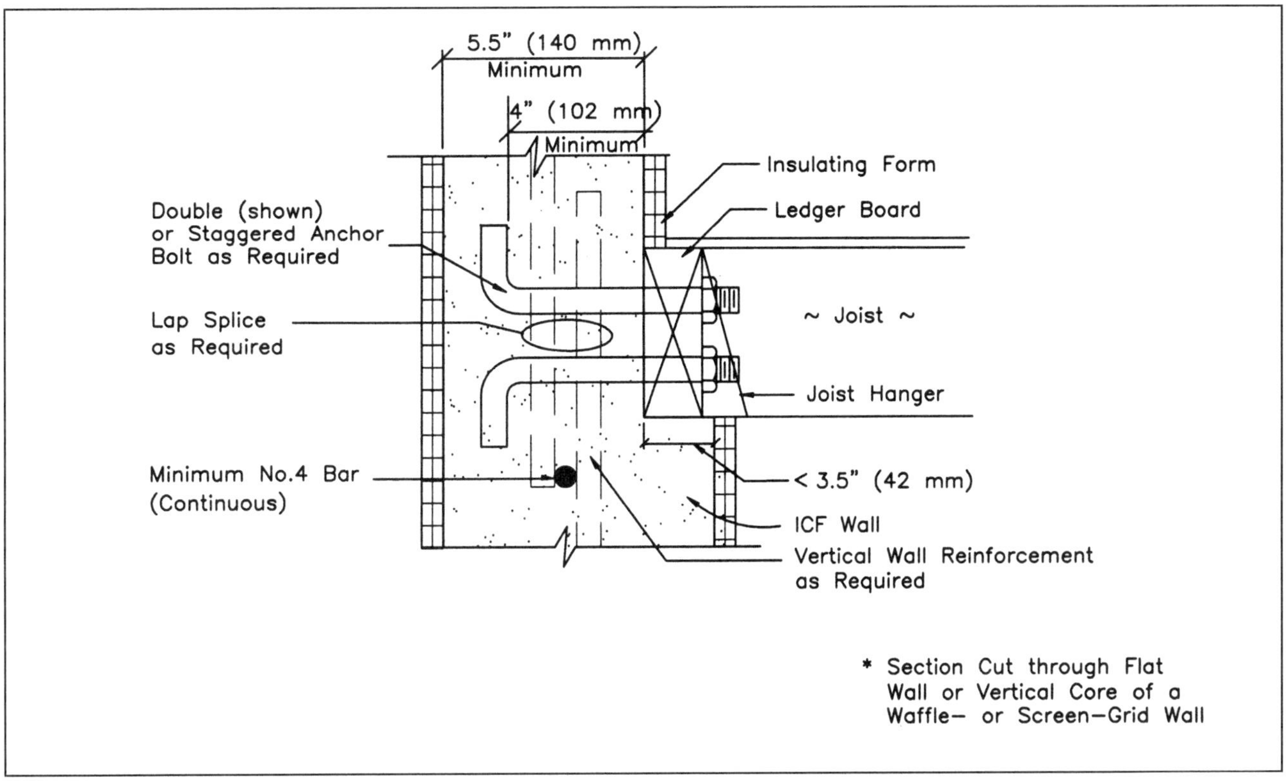

Figure 6.5 Floor Ledger-ICF Wall Connection (Side-Bearing Connection)

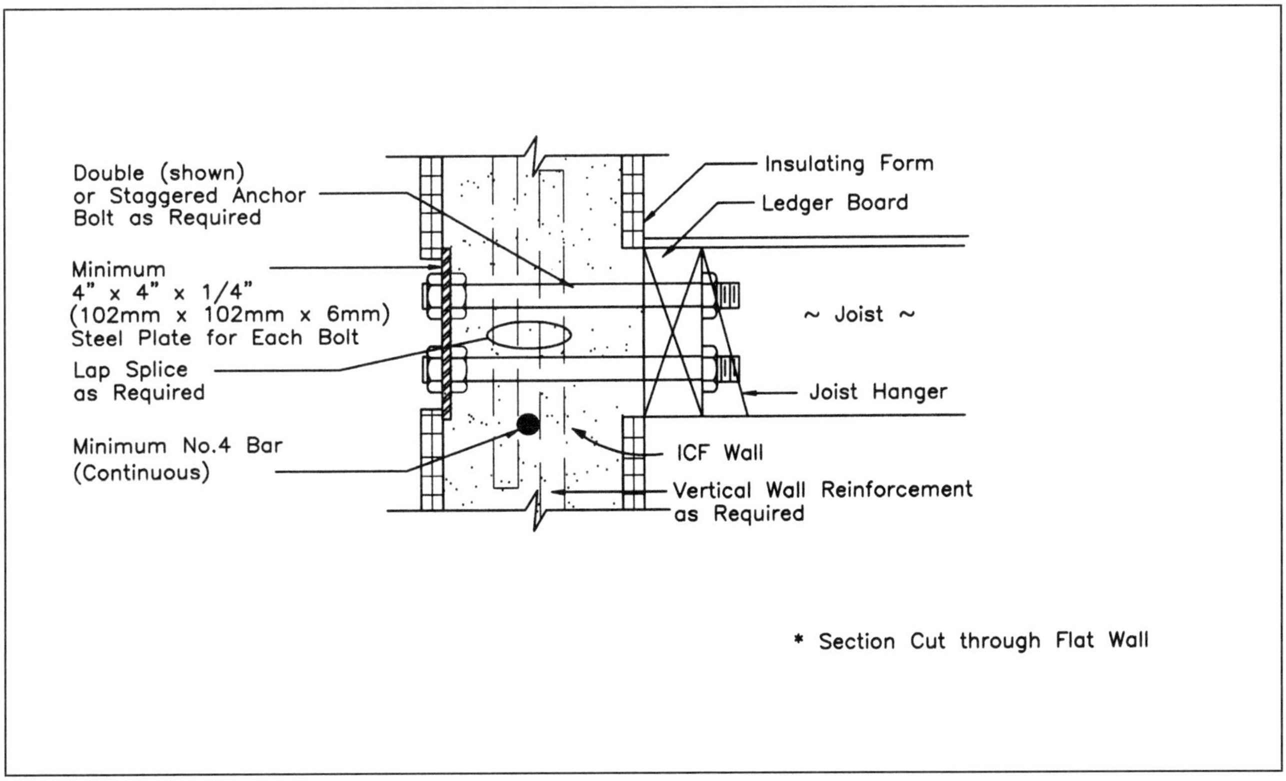

Figure 6.6 Floor Ledger-ICF Wall Connection (Through-Bolt Connection)

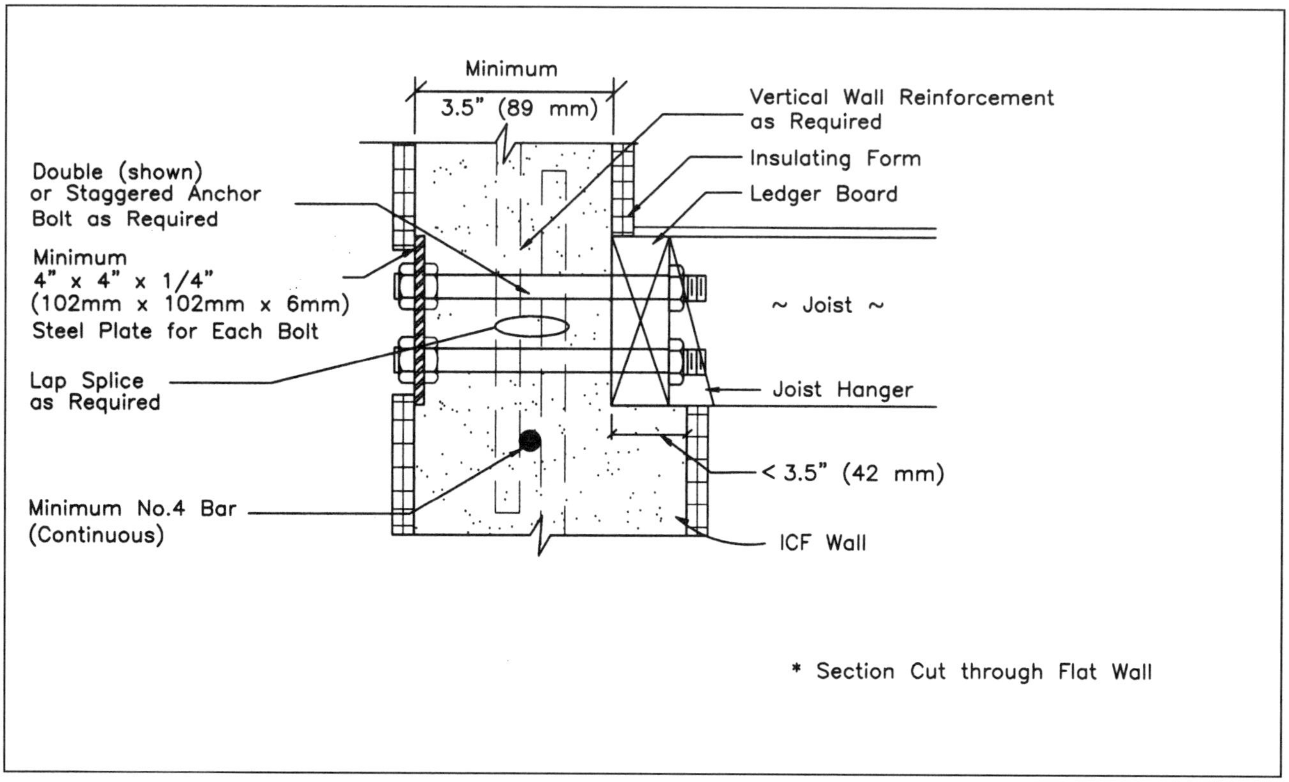

Figure 6.7 Floor Ledger-ICF Wall Connection (Through-Bolt Connection)

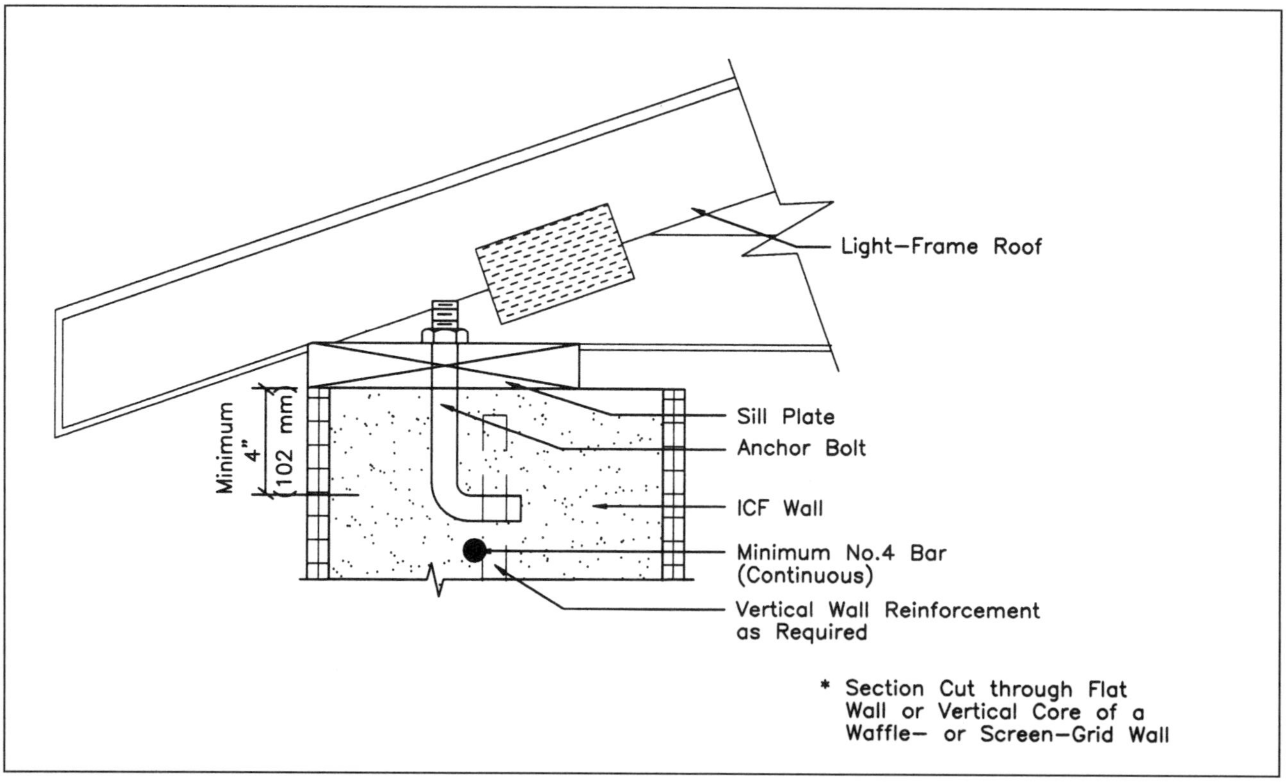

Figure 6.8 Top Wood Sill Plate-ICF Wall System Connection

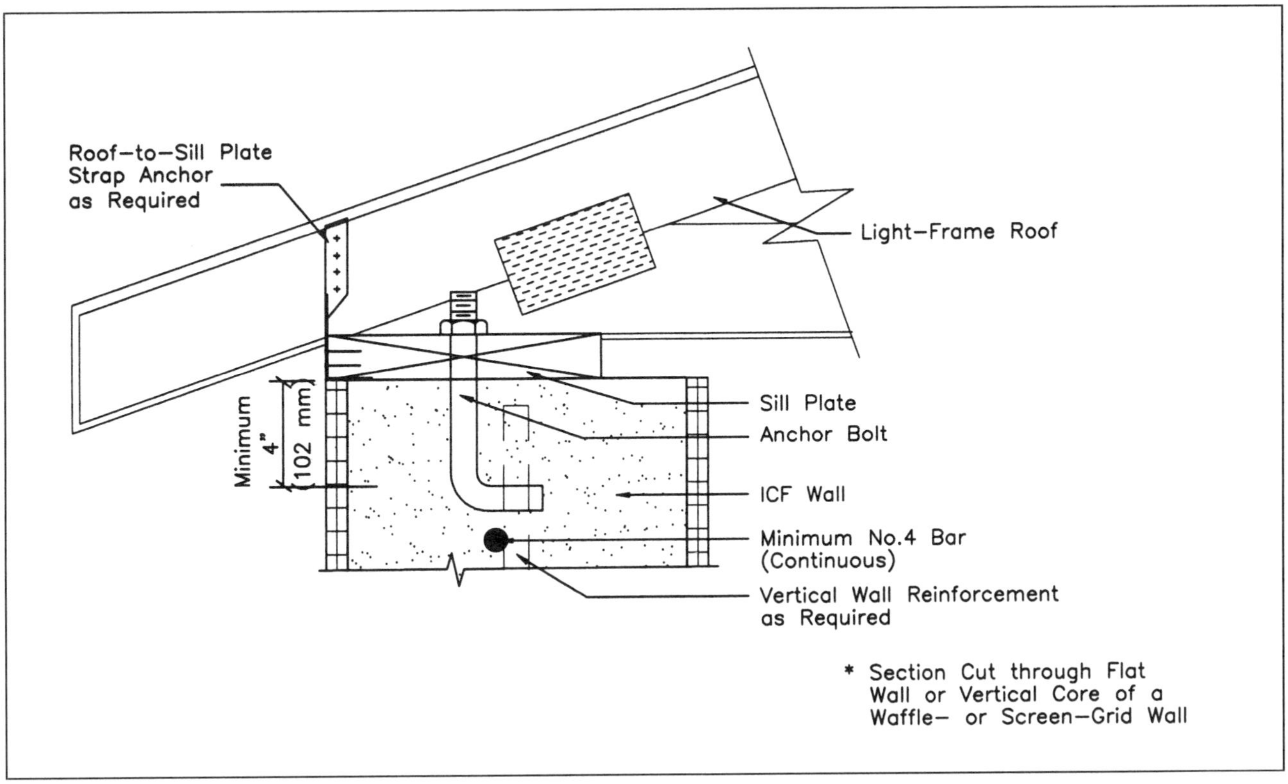

Figure 6.9 Roof-ICF Wall System Connection

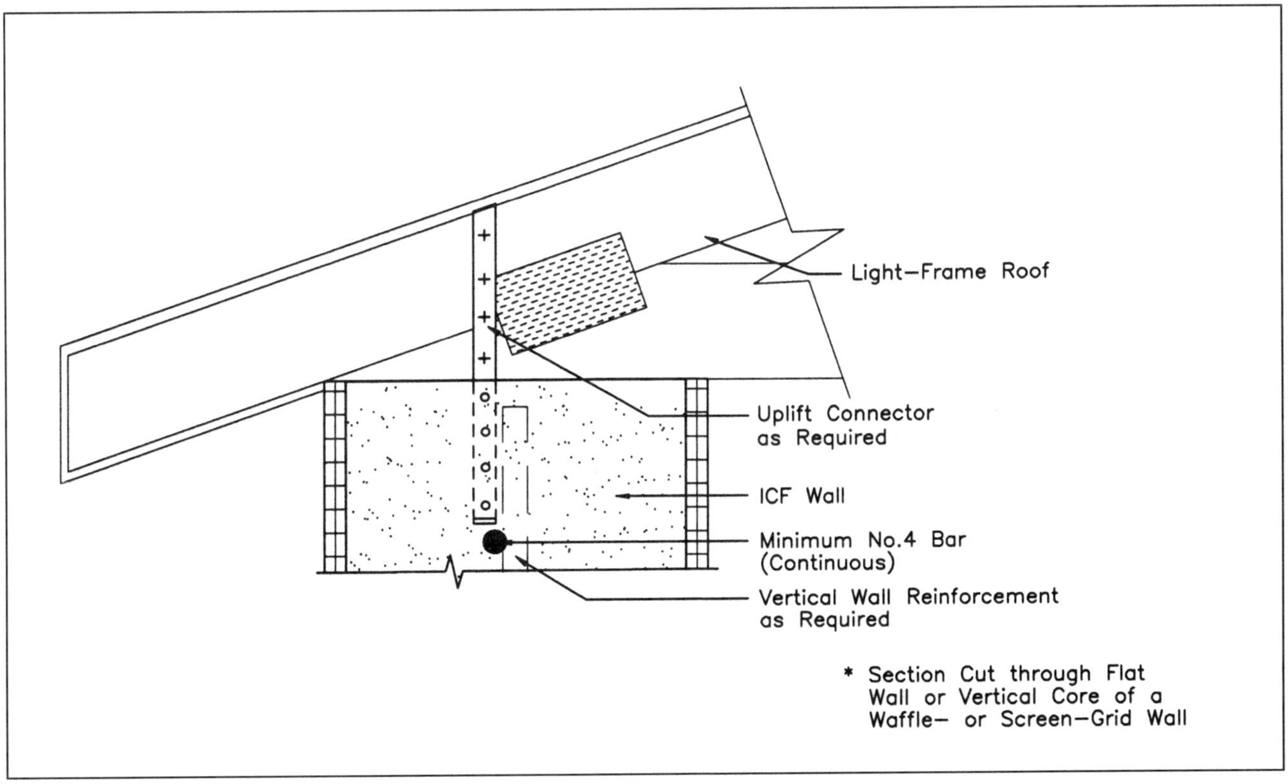

Figure 6.10 Roof-ICF Wall System Connection

7.0 UTILITIES

7.1 Plumbing Systems

Plumbing system installation shall comply with the applicable plumbing code.

7.2 HVAC Systems

HVAC system installation shall comply with the applicable mechanical code.

7.3 Electrical Systems

Electrical system installation shall comply with the National Electric Code.

8.0 CONSTRUCTION AND THERMAL GUIDELINES

8.1 Construction Guidelines

- Before placing concrete, formwork enclosures shall be cleaned of any debris and shall be free from frost. Concrete shall not be deposited into formwork containing snow, mud, or standing water or on or against any frozen material.

- An adequate method shall be followed to prevent freezing of concrete materials in cold-weather during the placement and curing process. The insulating form shall be considered as adequate protection against freezing when approved.

- Before placing concrete, vertical and horizontal reinforcement shall be secured in place within the insulating concrete form as required in Section 2.0. Concrete placing methods and equipment shall be such that the concrete is conveyed and deposited at the specified slump, without segregation and without significantly changing any of the other specified qualities of the concrete.

- Concrete shall be placed continuously, as close as possible to its final position. Concrete shall be deposited by using a chute, wheel barrow, crane and bucket, pump, belt conveyor, or other method that will not result in segregation of the concrete.

8.2 Thermal Guidelines

8.2.1 Energy Code Compliance

The insulation value (R-value) of all ICF wall systems shall meet or exceed the applicable provisions of the local energy code or the Model Energy Code [23].

8.2.2 Moisture

Form materials shall be protected against moisture intrusion through the use of approved exterior wall finishes in accordance with Sections 3.0 and 4.0.

8.2.3 Ventilation

ICF walls are generally more air-tight than typical light-frame wall construction. The natural ventilation rate shall not be less than that required by the local code or 0.35 ACH, whichever is greater. Mechanical ventilation shall be provided when required to meet the minimum air exchange rate of 0.35 ACH in accordance with the Model Energy Code [23].

9.0 REFERENCES

[1] ASTM E 380 *Standard Practice for Use of the International System of Units (SI) (the Modernized Metric System)*, American Society for Testing and Materials (ASTM), West Conshohocken, Pennsylvania. 1992.

[2] *Building Code Requirements for Structural Concrete* (ACI 318-95). American Concrete Institute, Detroit, Michigan. 1995.

[3] *Structural Design of Insulating Concrete Form Walls in Residential Construction.* Portland Cement Association, Skokie, Illinois. 1998.

[4] *Minimum Design Loads for Buildings and Other Structures (*ASCE 7-93). American Society of Civil Engineers, New York, New York. 1993.

[5] *Standard Building Code*. Southern Building Code Congress International, Inc. (SBCCI), Birmingham, Alabama. 1994.

[6] *Uniform Building Code*. International Conference of Building Officials (ICBO), Whittier, California. 1997.

[7] *National Building Code.* Building Officials & Code Administrators International, Inc. (BOCA), Country Club Hills, Illinois. 1993.

[8] *One- and Two-Family Dwelling Code, 1995 Edition*. The Council of American Building Officials (CABO), Falls Church, Virginia. 1995.

[9] *Guide to Residential Cast-in-Place Concrete Construction* (ACI 322R-84). American Concrete Institute, Detroit, Michigan. 1984.

[10] ASTM C 31/C 31M-96 *Standard Practice for Making and Curing Concrete Test Specimens in the Field.* American Society for Tcsting and Materials (ASTM), West Conshohocken, Pennsylvania. 1997.

[11] ASTM C 39-96 *Standard Test Method for Compressive Strength of Cylindrical Concrete Specimens*. American Society for Testing and Materials (ASTM), West Conshohocken, Pennsylvania. 1996.

[12] ASTM E 84-96a *Standard Test Method for Surface Burning Characteristics of Building Materials*. American Society for Testing and Materials (ASTM), West Conshohocken, Pennsylvania. 1996.

[13] ASTM C 143-90a *Standard Test Method for Slump of Hydraulic Cement.* American Society for Testing and Materials (ASTM), West Conshohocken, Pennsylvania. 1978.

[14] ASTM A 370-96 *Standard Test Methods and Definitions for Mechanical Testing of Steel Products,* American Society for Testing and Materials (ASTM), West Conshohocken, Pennsylvania. 1996.

[15] ASTM C 94-96e1 *Standard Specification for Ready-Mixed Concrete.* American Society for Testing and Materials (ASTM), West Conshohocken, Pennsylvania. 1996.

[16] ASTM A 615 / A 615 M-96a *Standard Specification for Deformed and Plain Billet-Steel Bars for Concrete Reinforcement.* American Society for Testing and Materials (ASTM), West Conshohocken, Pennsylvania. 1996.

[17] ASTM A 616 / A 616 M-96 *Standard Specification for Rail-Steel Deformed and Plain Bars for Concrete Reinforcement.* American Society for Testing and Materials (ASTM), West Conshohocken, Pennsylvania. 1996.

[18] ASTM A 617 / A 617 M-*96a Standard Specification for Axle-Steel Deformed and Plain Bars for Concrete Reinforcement,* American Society for Testing and Materials (ASTM), West Conshohocken, Pennsylvania. 1996.

[19] ASTM A 706 / A 706 M-96b *Standard Specification for Low-Alloy Steel Deformed and Plain Bars for Concrete Reinforcement.* American Society for Testing and Materials (ASTM), West Conshohocken, Pennsylvania. 1996.

[20] ASTM C 578-95 *Standard Specification for Rigid, Cellular Polystyrene Thermal Insulation,* American Society for Testing and Materials (ASTM). West Conshohocken, Pennsylvania. 1995.

[21] *Design Guide for Frost-Protected Shallow Foundations, Second Edition.* Prepared for the U.S. Department of Housing and Urban Development and the U.S. Department of Energy by the NAHB Research Center, Inc., Upper Marlboro, Maryland. 1996.

[22] ASTM E 119-95a *Standard Test Methods for Fire Tests of Building Construction and Materials.* American Society for Testing and Materials (ASTM), West Conshohocken, Pennsylvania. 1995.

[23] *Model Energy Code.* The Council of American Building Officials (CABO), Falls Church, Virginia. 1995.

PART II

COMMENTARY

INTRODUCTION

The *Commentary* is provided to facilitate the use of, and provide background information for, the *Prescriptive Method.* It also includes supplemental information and engineering assumptions as well as the methods that underpin the *Prescriptive Method.* It presents the individual sections, figures, and tables in the same sequence found in the *Prescriptive Method.* For detailed engineering calculations, refer to Appendix B, *Engineering Technical Substantiation.*

Information is presented in both U.S. customary units and International System (SI) units except for reinforcement bar sizes which are only presented in U.S. customary units. Refer to Appendix C for the corresponding reinforcement bar size in SI units.

C1.0 GENERAL

C1.1 Purpose

The goal of the *Prescriptive Method* is to present prescriptive criteria (i.e., tables, figures, guidelines) for the construction of one- and two-story dwellings constructed with insulating concrete forms. Before development of this document, no prescriptive standards were available to builders and code officials for the purpose of constructing insulating concrete form houses without the added expense of a design professional and the other costs associated with using a "nonstandard" material for residential construction.

The *Prescriptive Method* presents the minimum requirements for basic residential construction that is consistent with the safety levels specified in the current U.S. building codes governing residential construction.

The *Prescriptive Method* is not applicable to all possible conditions of use and is subject to the applicability limits set forth in Table 1.1 of the *Prescriptive Method.* The applicability limits should be carefully understood as they define important constraints on the use of the *Prescriptive Method.* This document is not intended to restrict the use of either sound judgment or exact engineering analysis of specific applications that may result in improved designs and economy.

C1.2 Approach

The requirements, figures, and tables provided in the *Prescriptive Method* are based primarily on the *Building Code Requirements for Structural Concrete* [C1] and the *Structural Design of Insulating Concrete Form Walls in Residential Construction* [C2], and the pertinent requirements of the *Minimum Design Loads for Buildings and Other Structures* [C3], the *One- and Two-Family Dwelling Code* [C4], the *National Building Code* [C5], the *Uniform Building Code* [C6], and the *Standard Building Code* [C7]. Construction practices from the *Guide to Residential Cast-in-Place Concrete Construction* [C8] have also been used. Engineering decisions requiring interpretations or judgments in applying the above references are documented in this *Commentary.*

C1.3 Scope

It is unrealistic to develop an easy-to-use document that provides prescriptive requirements for all types and styles of ICF construction. Therefore, the *Prescriptive Method* is limited in its applicability to typical one- and two-family dwellings. The requirements set forth in the *Prescriptive Method* apply only to the construction of ICF houses that meet the limits set forth in Table 1.1 of the *Prescriptive Method.* The applicability limits are necessary for defining reasonable boundaries to the conditions that must be considered in developing prescriptive construction requirements. The *Prescriptive Method,* however, does not limit the application of alternative methods or materials through engineering design by a design professional.

The basic applicability limits are based on industry convention and experience. Detailed applicability limits were documented in the process of developing prescriptive design requirements for various elements of the structure. In some cases, engineering sensitivity analyses were performed to help define appropriate limits.

The applicability limits strike a reasonable balance among engineering theory, available test data, and proven field practices for typical residential construction applications. They are intended to prevent misapplication while addressing a reasonably large percentage of new housing conditions. Special consideration is directed toward the following items related to the applicability limits.

Building Geometry

The provisions in the *Prescriptive Method* apply to detached one- or two-family dwellings, townhouses, and other attached single-family dwellings not more than two stories in height above grade. Given the maximum building plan dimensions and the roof and floor clear spans specified in Table 1.1 of the *Prescriptive Method*, the largest rectangular home that can be constructed without a center load-bearing wall or beam is a home 40 feet by 60 feet (12 m by 18 m). The largest rectangular home that can be constructed with a center load-bearing wall or beam is a home 60 feet by 60 feet (18 m by 18 m) provided that the roof and each floor is supported by a center load-bearing wall or beam to reduce floor and roof clear spans. Application to homes with complex architectural configurations is subject to careful interpretation and sound judgment by the user; design support may be required.

Site Conditions

Snow loads are typically given in a ground snow load map such as that provided in ASCE 7 [C3] or by local practice. The 0 to 70 psf (0 to 3.4 kPa) ground snow load used in the *Prescriptive Method* covers approximately 90 percent of the United States, which includes the majority of the houses that are expected to use this document. Houses in areas with higher ground snow loads cannot use this document and should consult a design professional.

All areas of the United States fall within the 70 to 110 mph (113 to 177 km/hr) (fastest-mile) range of design wind speeds [C4][C5][C6][C7]. Houses built along the immediate, hurricane-prone coastline subjected to storm surge (i.e., beach-front property) cannot use this document; a design professional should be consulted. The National Flood Insurance Program (NFIP) requirements, administered by the Federal Emergency Management Agency (FEMA), should also be employed for structures located in coastal high-hazard zones.

Buildings constructed in accordance with the *Prescriptive Method* are limited to regions designated as Seismic Zones 0, 1, and 2 [C4][C6]. The design of buildings in Seismic Zones 3 and higher [C4][C6] are presently beyond the scope of the engineering provisions of this document and require the services of a design professional.

Soil borings are rarely required for residential construction except where there are known risks or a history of problems (i.e., organic deposits, landfills, expansive soils) associated with building in certain areas. The presumptive soil-bearing value of 2,000 psf (96 kPa) is based on typical soil conditions in the United States except in areas of high risk or where geotechnical investigation proves otherwise.

Loads

Building codes and standards handle loads and load combinations differently. Consistent values were established for design loads in accordance with a review of the major building codes and standards. The results of this load review are embodied in the applicability limits table in the *Prescriptive Method.* Loads and load combinations requiring calculations to analyze the structural components and assemblies of a home are presented in Appendix B, *Engineering Technical Substantiation.*

C1.4 ICF System Limitations

All ICF systems are typically categorized with respect to the form itself and the resulting shape of the formed concrete wall. There are three types of ICF forms: panel, plank, and block. The differences among the ICF form types are their size and attachment requirements. The different form types exist mainly for ease of installation based on use, available resources, and builder preferences and do not necessarily affect the structural capacity of the wall.

There are also three categories of ICF systems based on the resulting shape of the formed concrete wall. From a structural design standpoint, it is only the shape of the concrete inside the form, not the type of ICF form, that is of importance. The shape of the concrete wall may be better understood by visualizing the form stripped away from the concrete, thereby exposing it to view. The three categories of ICF wall form are flat, grid, and post-and-beam. The grid wall type is further categorized into waffle-grid and screen-grid wall systems. These classifications are provided solely to ensure that the design tables in this document are applied to the ICF wall systems as the authors intended.

The post-and-beam ICF wall system is not included in this document because it requires a different engineering analysis. It is analyzed as a concrete frame rather than as a monolithic concrete (i.e., flat, waffle-grid, or screen-grid) wall construction in accordance with ACI 318 [C1]. Post-and-beam systems may be analyzed in the near future to provide a prescriptive method for post-and-beam ICF wall systems.

C1.5 Definitions

The definitions in the *Prescriptive Method* are provided because certain terms are likely to be unfamiliar to the home building trade; the definitions are provided to clarify and define the terminology. Additional definitions that warrant technical explanation are defined below.

Permeance: The permeability of a porous material; a measure of the ability of moisture to migrate through a material.

Superplasticizer: A substance added to concrete mix that improves workability at very low water-cement ratios to produce high, early-strength concrete. Also referred to as high-range, water-reducing admixtures.

C2.0 MATERIALS, SHAPES, AND STANDARD SIZES

C2.1 Physical Dimensions

Due to industry variations related to the dimensions of ICFs, dimensions were standardized (i.e., thickness, width, spacing) to allow for the development of the *Prescriptive Method.* This prescriptive approach may result in a conservative design for ICFs where thickness and width are greater than the minimum allowable or the spacing of vertical cores is less than the maximum allowable. Consult a design professional if a more economical design is desired.

C2.1.1 Flat ICF Wall Systems

Wall Thickness: The actual wall thickness is limited to 3.5 inches (89 mm), 5.5 inches (140 mm), 7.5 inches (191 mm), or 9.5 inches (241 mm) in order to accommodate ICF flat wall systems currently available. ICF flat wall manufacturers whose products have a wall thickness greater than those listed above shall use the tables in the *Prescriptive Method* for the nearest available wall thickness that does not exceed the actual wall thickness.

C2.1.2 Waffle-Grid ICF Wall Systems

Core Thickness and Width: The vertical and horizontal core thickness and width are limited per Table 2.1 in the *Prescriptive Method* in order to accommodate ICF waffle-grid wall systems currently available. Variation among the ICF waffle-grid manufacturers is minimal; therefore, the tables in the *Prescriptive Method* should produce economical designs for a maximum 32-foot- (9.8-m-) wide building meeting the applicability limits of Table 1.1 in the *Prescriptive Method.* ICF waffle-grid manufacturers that offer concrete cross sections larger than those required in Table 2.1 of the *Prescriptive Method* shall use the tables for the nominal size that has the nearest available core thickness not exceeding the actual wall thickness. Although Figure 2.2 in the *Prescriptive Method* shows the ICF waffle-grid vertical core shape as elliptical, the shape of the vertical core may be round, square, or rectangular provided that the minimum dimensions in Table 2.1 are met.

Core Spacing: The vertical and horizontal core spacing is limited per Table 2.1 of the *Prescriptive Method* in order to accommodate the ICF waffle-grid wall systems currently available. Variation in the products offered by the ICF waffle-grid manufacturers is minimal; therefore, the tables in the *Prescriptive Method* should produce economical designs for a maximum 32-foot- (9.8-m-) wide building meeting the applicability limits of Table 1.1 in the *Prescriptive Method.*

Web Thickness: The minimum web thickness of 2 inches (51 mm) is based on ICF waffle-grid systems currently available. Variation in the products offered by the ICF waffle-grid manufacturers is minimal; therefore, the tables in the *Prescriptive Method* should produce economical designs for a maximum 32-foot- (9.8-m-) wide building meeting the applicability limits of Table 1.1 in the *Prescriptive Method.*

C2.1.3 Screen-Grid ICF Wall System

Core Thickness and Width: The vertical and horizontal core thickness and width are limited per Table 2.2 in the *Prescriptive Method* in order to accommodate ICF screen-grid wall systems currently available. ICF screen-grid manufacturers that offer concrete cross sections larger than those required in Table 2.2 shall use the tables for the nominal size that has the nearest available core thickness not exceeding the actual wall thickness. Although Figure 2.3 of the *Prescriptive Method* shows the ICF screen-grid vertical core shape as round, the shape of the vertical core may be square, rectangular, elliptical, or other shape provided that the minimum dimensions in Table 2.2 are met.

Core Spacing: The vertical and horizontal core spacing is limited per Table 2.2 of the *Prescriptive Method* in order to accommodate the large number of ICF screen-grid wall systems currently available. Due to a lack of test data to suggest otherwise, the maximum allowable horizontal and vertical core spacing is a value agreed on by the steering committee members. The core spacing is the main requirement differentiating an ICF screen-grid system from an ICF post-and-beam system. Future testing is required to determine the maximum allowable core spacing without adversely affecting the wall system's ability to act as a wall rather than as a frame.

C2.2 Concrete Materials

C2.2.1 Concrete Mix

The maximum slump and aggregate size requirements are based on current ICF practice. Considerations included in the prescribed maximums are ease of placement, ability to fill cavities thoroughly, and limiting the pressures exerted on the form by wet concrete.

Concrete for walls less than 8 inches (203 mm) thick is typically poured into the forms by using a 2-inch- (51-mm-) diameter boom or line pump; aggregates larger than the maximums prescribed may clog the line. The ability to fill cavities during concrete placement for ICF flat wall systems depends little on slump and aggregate size because the only obstructions during the pour are the form ties around which the concrete readily flows. In contrast, the ability to fill cavities during concrete placement for ICF waffle- and screen-grid wall systems is more dependent on slump and aggregate size due to the presence of the form ties and, more importantly, the horizontal and vertical cavities, which may encourage the formation of honeycombs and voids if the concrete is not properly placed.

To determine the most effective mix, the industry is currently conducting experiments that vary slump and aggregate size, and use admixtures (i.e., superplasticizers). The research may not produce an industrywide standard due to the variety of available form material densities and ICF types; therefore, an exception for higher allowable slumps is provided in the *Prescriptive Method.*

C2.2.2 Compressive Strength

The minimum concrete compressive strength requirement is based on the minimum current ICF practice. This prescriptive approach may result in a conservative design for ICFs where greater concrete compressive strength is required by the ICF manufacturer. Consult a design professional if a more economical design is desired.

It is believed that concrete cured in ICFs produces higher strengths than conventional concrete construction because the formwork is never intended to be removed; however, the concrete compressive strength specified herein is based on cylinder tests cured outside the ICF in accordance with ASTM C31 [C9] and ASTM C 39 [C10].

C2.2.3 Reinforcing Steel

Materials: The *Prescriptive Method* applies to steels with minimum yield strengths of 40 ksi (300 MPa). This prescriptive approach may result in a conservative design for ICFs where greater reinforcement yield strength is required by the ICF manufacturer. Consult a design professional if a more economical design is desired.

It is believed that Grade 40 (300 MPa) is no longer widely available. Tables using Grade 60 (420 MPa) steel reinforcement would prove beneficial. Tables using materials other than steel may also prove beneficial; however, testing may be required.

Placement: The *Prescriptive Method* requires vertical and horizontal wall reinforcement to be placed in the middle third of the wall thickness. The requirements for vertical and horizontal wall reinforcement placement are based on current construction practice for a large number of ICF manufacturers. They provide clearance from the center of the wall on which the calculations are based for reinforcement lap splices and intersections of horizontal and vertical wall reinforcement.

A few ICF manufacturers produce a groove or loop in the form tie allowing for easier reinforcement placement. These manufacturers may locate the groove or loop closer to the interior or exterior face of the wall to reap the maximum benefit from the steel reinforcement; the location depends on the wall's loading conditions and is reflected in the exception for basement walls as well as in the middle-third requirement for above-grade walls.

Lap splices are provided to transfer forces from one bar to another where continuous reinforcement is not practical. Lap splices are typically required between wall stories, at building corners, and for continuous horizontal wall reinforcement. The lap splice requirements are based on ACI 318 [C1] and may be considered excessive for the loads experienced in residential construction. Tests may prove beneficial to determine allowable reductions in lap splice lengths for use in residential construction.

C2.3 Form Materials

The materials listed in the *Prescriptive Method* are based on currently available ICFs. From a structural standpoint, the material can be anything that has sufficient strength to contain the concrete during placement and curing. From a thermal standpoint, the form material should provide the R-value required by the local building code; however, the required R-value could be met by installing additional insulation to the exterior or interior of the form, provided that it does not reduce the minimum concrete dimensions as specified in Section 2.0. From a life-safety standpoint, the form material can be anything that meets the criteria for flame-spread and smoke development. The *Prescriptive Method* addresses other concerns (i.e., water vapor transmission, termite resistance) that must be considered when using materials other than those specifically listed here. This section is not intended to exclude the use of either a current or future material provided that the requirements of this document are met.

C3.0 FOUNDATIONS

C3.1 Footings

The loads imposed on the footings do not vary from those of conventional concrete construction; however, the *Prescriptive Method* provides a table for minimum footing widths with ICF construction. ICF footing forms are currently available and may be used if they meet the minimum footing dimensions required in Table 3.1 in the *Prescriptive Method*. Table 3.1 is similar to the requirements in CABO [C4] for 8-inch- (203-mm-) solid or fully grouted masonry. The minimum footing width values are based on a 28-foot- (8.5-m-) wide building.

Minimum footing widths are based on the maximum loading conditions found in Table 1.1 of the *Prescriptive Method* a minimum footing depth of 12 inches (305 mm) below grade, story heights of 8 feet (2.4 m), 9 feet (2.7 m), and 10 feet (3 m), and the assumption that all stories are the same thickness and are constructed of ICFs unless otherwise noted.

The values in Table 3.1 of the *Prescriptive Method* for a one-story ICF structure account for one ICF story above-grade. The values in Table 3.1 for a two-story ICF structure account for either one ICF story above-grade with an ICF basement or two ICF stories above-grade. The values in the table for a three-story ICF structure account for two ICF stories above-grade with an ICF basement. The values in the table assume the wall width of all stories is uniform unless noted otherwise.

Footnote 1 to Table 3.1 in the *Prescriptive Method* provides guidance for sizing an unreinforced footing based on rule of thumb. This requirement may be relaxed when the footing is reinforced or a design professional is consulted. Soil borings are rarely required for residential construction except where there are known risks or a history of problems (i.e., organic deposits, landfills, expansive soils) associated with building in certain areas. For an approximate relationship between soil type and load-bearing value, refer to Table C3.1.

C3.2 ICF Foundation Wall Requirements

The *Prescriptive Method* provides reinforcement tables for foundation walls constructed within the applicability limits of Table 1.1 in the *Prescriptive Method*. The maximum design conditions are Seismic Zone 2, ground snow load of 70 psf (3.4 kPa), and equivalent fluid density of 60 pcf (960 kg/m^3). The *Prescriptive Method* provides the minimum required vertical and horizontal wall reinforcement for various equivalent fluid densities, wall heights, and unbalanced backfill heights. Vertical wall reinforcement tables are limited to foundation walls with load-bearing wall heights up to 10 feet (3 m).

Residential construction makes widespread use of 8-foot (2.4-m) walls; however, ICF homes are often constructed with higher ceilings. Walls are grouped into three categories as follows:

- walls with soil backfill having a maximum 30 pcf (481 kg/m^3) equivalent fluid density;

- walls with soil backfill having a maximum 45 pcf (721 kg/m^3) equivalent fluid density;
- walls with soil backfill having a maximum 60 pcf (960 kg/m^3) equivalent fluid density.

The following design assumptions were made in analyzing the walls:

- Walls support either one or two stories above.
- Walls are simply supported for each story.
- Walls contain no openings.
- Bracing is provided for the wall by the floors above and floor slabs.
- Roof slopes range from 0:12 to 12:12.
- Deflection criterion is the height of the wall, in inches, divided by 240.
- Ceilings, roofs, attics, and floors span the full width of the house (assume no interior load-bearing walls or beams).
- Roof snow loads were calculated by multiplying the ground snow load by 0.7. Therefore, the roof snow load was taken as $P=0.7P_g$, where P_g is the ground snow load in pounds per square foot.

Deflection limits are primarily established with regard to serviceability concerns. The intent is to prevent excessive deflection, which may result in cracking of finishes. For walls, most codes generally agree that *L/240* represents an acceptable serviceability limit for deflection. For walls with flexible finishes, less stringent deflection limits may be used. The reader is referred to Appendix B, *Engineering Technical Substantiation* for an example calculation for a foundation wall. In cases where the calculations required no vertical wall reinforcement, the following minimum wall reinforcement is required.

- A minimum of one vertical No. 4 bar at 48 inches (1.2 m) on center for 5.5-inch (140-mm) thick foundation walls to account for temperature, shrinkage, potential honeycombing, voids, or construction errors.

Minimum horizontal wall reinforcement is based on recommendations in *Design Criteria for Insulating Concrete Form Wall Systems* [C11]. The minimum allows for temperature, shrinkage, potential honeycombing, voids, or construction errors.

C3.2.1 ICF Walls with Slab-on-Grade

ICF stem wall thickness and height are determined as those which can distribute the building loads safely to the earth. The stem wall thickness should be greater than or equal to the wall thickness for the above-grade wall it supports. Given that stem walls are relatively short and are backfilled on both sides of the wall, earth loads induce a small amount of bending moment on the walls; accordingly, lateral bracing should not be required before backfilling.

C3.2.2 ICF Crawlspace Walls

Table 3.2 in the *Prescriptive Method* applies to crawlspace walls 5 feet (1.5 m) or less in height with a maximum unbalanced backfill height of 4 feet (1.2 m). These values were derived from the *Structural Design of Insulating Concrete Form Walls in Residential Construction* [C2]. Loading conditions were based on a maximum 32-foot- (9.8-m-) wide building with the lightest and heaviest practical gravity loads experienced in residential construction. The values for minimum vertical wall reinforcement are based on the controlling loading condition. For detailed engineering calculations, refer to Appendix B, *Engineering Technical Substantiation.*

Soil borings are rarely required for residential construction except where there are known risks or a history of problems (i.e., organic deposits, landfills, expansive soils) associated with building in certain areas. Refer to Table C3.2 for an approximate relationship between soil classifications and equivalent fluid density [C3].

Backfilling is not to occur without lateral support at the top of the wall from either the ground floor structure or temporary bracing unless the backfill height is less than one-half the crawlspace wall height. This requirement ensures that the backfill does not cause the wall to overturn. Concrete walls can withstand higher lateral load created from the backfill when the top of the wall is braced and axial loads are present on the wall. Typically, providing lateral bracing at the top of the wall until the structure above is in place is sufficient. Moreover, backfilling should not occur before seven days after the concrete pour when using a minimum concrete compressive strength of 2,500 psi (17 MPa); waiting seven days typically allows the concrete to reach sufficient strength.

C3.2.3 ICF Basement Walls

Tables 3.3 through 3.9 in the *Prescriptive Method* pertain to basement walls. The values were derived from the *Structural Design of Insulating Concrete Form Walls in Residential Construction* [C2]. Loading conditions were based on a maximum 32-foot- (9.8-m-) wide building with the lightest and heaviest practical gravity loads experienced in residential construction. The values for minimum vertical wall reinforcement are based on the controlling loading condition. For detailed engineering calculations, refer to the Appendix B, *Engineering Technical Substantiation.*

Soil borings are rarely required for residential construction except where there are known risks or a history of problems (i.e., organic deposits, landfills, expansive soils) associated with building in certain areas. Refer to Table C3.2 for an approximate relationship between soil classifications and equivalent fluid density.

Backfilling is not to occur without lateral support at the top of the wall from either the ground floor structure or temporary bracing unless the unbalanced backfill height is less than one-half the basement wall height. This requirement ensures that the backfill does not cause the wall to overturn. Concrete walls can withstand higher lateral loads created from the backfill when the top of the wall is braced and axial loads are present on the wall. Typically, providing lateral bracing at the top of the wall until the structure above is in place is sufficient. Moreover, backfilling should not occur before seven days after the concrete pour when using a minimum concrete compressive strength of 2,500 psi (17 MPa); waiting seven days typically allows the concrete to reach sufficient strength.

C3.3 ICF Foundation Wall Coverings

The requirements for interior covering of habitable spaces are based on current building codes and are self-explanatory.

It is generally accepted that a monolithic concrete wall is a solid wall through which water and air cannot readily flow; however, there is a possibility that the concrete wall may have honeycombs, voids, or hairline cracks through which water may enter. Voids in concrete are inherent in screen-grid ICF walls and will allow water to enter the structure. As a result, a moisture barrier on the exterior face of all ICF below-grade walls is generally required and should be considered good practice. Due to the variety of materials on the market, waterpproofing and dampproofing materials are typically specified by the ICF manufacturer. The limitation in the *Prescriptive Method* regarding nonpetroleum-based materials reflects the concern that many ICFs are currently manufactured of rigid foam plastic, which is generally incompatible with petroleum-based materials.

A vapor retarder may be required on the interior face of the ICF wall in some cases. Test results have shown a potential exists for condensation occurring on the interior face of above-grade ICFs with a permeance as little as 0.5 perms in colder climates. Few problems have been reported when the exterior wall finishes are properly designed and constructed to prevent water intrusion. The reader is referred to *Mitigation of Moisture in Insulating Concrete Form Wall Systems* [C12] for more information on the testing and suggested construction recommendations.

C3.4 Termite Protection Requirements

Termites need wood (cellulose) and moisture to survive. Rigid foam plastic provides termites with no nutrition but can provide access to the structure. Recently, some building codes have prohibited rigid foam plastics for near- or below-grade use in heavy termite infestation areas. Code officials and termite treaters fear that foam insulation provides a "hidden pathway". Local building code requirements, the local pest control company, and the ICF manufacturer should be consulted regarding this concern to determine appropriate solutions when required A brief list of some possible termite control measures follow; however, research is currently underway to determine other alternatives.

- Rely on soil treatment as a primary defense against termites. Periodic retreatment and inspection should be carried forth by the home owner.
- Install termite shields.
- Provide a 6-inch- (152-mm-) high clearance above finish grade around the perimeter base of the structure where the foam has been removed to allow visual detection of termites.

TABLE C3.1
LOAD-BEARING SOIL CLASSIFICATION

Minimum Load-Bearing Value psf (kPa)	Soil Description
2,000 (96)	Clay, sandy clay, silty clay, and clayey silt
3,000 (144)	Sand, silty sand, clayey sand, silty gravel, and clayey gravel
4,000 (192)	Sandy gravel and medium-stiff clay
> 4,000 (192)	Stiff clay, gravel, sand, sedimentary rock, and crystalline bedrock

TABLE C3.2
EQUIVALENT FLUID DENSITY
SOIL CLASSIFICATION

Maximum Equivalent Fluid Density pcf (kg/m^3)	UCS[1] Classification	Soil Description
30 (481)	GW, GP, SW, SP, GM	Well-drained, cohesionless soils such as clean (few or no fines) sand and gravels
45 (721)	GC, SM	Well-drained, cohesionless soils such as sand and gravels containing silt or clay
60 (961)	SC, MH, CL, CH, ML-CL	Well-drained, inorganic silts and clays that are broken up into small pieces

1 UCS - Uniform Soil Classification system

C4.0 ICF ABOVE-GRADE WALLS

C4.1 ICF Above-Grade Wall Requirements

The *Prescriptive Method* provides reinforcement tables for walls constructed above-grade within the applicability limits of Table 1.1 in the *Prescriptive Method.* The maximum design conditions are Seismic Zone 2, ground snow load of 70 psf (3.4 kPa), and design (fastest-mile) wind speed of 110 mph (177 km/hr). The *Prescriptive Method* provides the minimum required vertical and horizontal wall reinforcement for different wind speeds, wall heights, and ground snow loads. Vertical wall reinforcement tables are limited to one- and two-story buildings with load-bearing wall heights up to 10 feet (3 m) per story.

Residential construction makes widespread use of 8-foot (2.4-m) walls; however, ICF homes are often constructed with higher ceilings. Walls are grouped into three categories as follows:

- walls for one-story or the second floor of a two-story building (supporting light-frame roof only);
- walls for the first story of a two-story building where the second story is light-frame construction (supporting light-frame second story and roof); and
- walls for the first story of a two-story building where the second story is ICF construction (supporting ICF second story and light-frame roof).

The following design assumptions were made in analyzing the walls:

- Walls are simply supported for each story.
- Walls contain no openings.
- Bracing is provided for the wall by the floors, slabs, and roof.
- Roof slopes range from 0:12 to 12:12.
- Deflection criterion is the height of the wall per story, in inches, divided by 240.
- Ceilings, roofs, attics, and floors span the full width of the house (assume no interior load-bearing walls or beams).
- Wind loads were calculated in accordance with SBC [C7] using components and cladding coefficients for enclosed buildings, interior zone, and mean roof height of 35 feet (11 m).
- Roof snow loads were calculated by multiplying the ground snow load by 0.7. Therefore, the roof snow load was taken as $P=0.7P_g$, where P_g is the ground snow load in pounds per square foot.

Deflection limits are primarily established with regard to serviceability concerns. The intent is to prevent excessive deflection, which may result in cracking of finishes. For walls, most codes generally agree that *L/240* represents an acceptable serviceability limit for deflection. For walls with flexible finishes, less stringent deflection limits may be used. The reader is referred to Appendix B, *Engineering Technical Substantiation* for an example calculation for an above-grade wall. In cases where the calculations required no vertical wall reinforcement, the following minimum wall reinforcement is required.

- A minimum of one vertical No. 4 bar at 32 inches (813 mm) on center for 3.5-inch (89-mm) thick above-grade walls to account for temperature, shrinkage, potential honeycombing, voids, or construction errors

- In conditions with wind speeds 90 mph (145 km/hr) or greater, a minimum of one vertical No. 4 bar at 8 feet (2.4 m) on center based on current masonry construction practice and past successful experience

Minimum horizontal wall reinforcement is based on recommendations in *Design Criteria for Insulating Concrete Form Wall Systems* [C11]. The minimum allows for temperature, shrinkage, or potential construction errors.

The more stringent requirement that vertical wall reinforcement be terminated with a bend or hook in high wind areas is based on current standards for masonry construction. The requirement has proven very effective in masonry construction in conditions with wind speeds 90 mph (145 km/hr) or greater. The bend or hook provides additional tensile strength in the concrete wall to resist the large roof uplift loads in high wind areas. The length of bend or hook may be considered excessive for concrete residential construction; tests may prove beneficial to determine allowable reductions in bend or hook lengths required.

C4.2 ICF Above-Grade Wall Coverings

The requirements for interior covering of habitable spaces are based on current building codes and are self-explanatory.

It is generally accepted that a monolithic concrete wall is a solid wall through which water and air cannot readily flow; however, there is a possibility that the concrete wall may have honeycombs, voids, or hairline cracks through which water may enter. Voids in concrete are inherent in screen-grid ICF walls and will allow water to enter the structure. As a result, a moisture barrier on the exterior face of the ICF wall is generally required and should be considered good practice.

A vapor retarder may also be required on the interior face of the ICF wall in some cases. Test results have shown a potential exists for condensation occurring on the interior face of above-grade ICFs with a permeance as little as 0.5 perms in colder climates. Few problems have been reported when the exterior wall finishes are properly designed and constructed to prevent water intrusion. The reader is referred to *Mitigation of Moisture in Insulating Concrete Form Wall Systems* [C12] for more information on the testing and suggested construction recommendations.

C5.0 ICF WALL OPENING REQUIREMENTS

C5.1 Minimum Length of ICF Wall without Openings

The tables in Sections 3.0 and 4.0 are based on ICF walls without door or window openings. This simplified approach rarely arises in residential construction since walls generally contain windows and doors. The amount of openings affects the lateral (racking) strength of the building, particularly for wind and seismic loading conditions. The *Prescriptive Method* provides recommendations for the amount and placement location of additional reinforcement required around openings. It also addresses the minimum amount of solid wall required to resist racking loads from wind and moderate seismic forces (i.e., Seismic Zones 0, 1, and 2).

The values for the base percentage of solid wall length along exterior wall lines listed in Table 5.3 of the *Prescriptive Method* were calculated using the main wind force resisting wind loads in accordance with SBC [C7] and seismic loads in accordance with ASCE 7 [C3]. The ICF wall was checked using resistance models for a building 30 feet by 60 feet (9 m by 18 m).

A shear model assumed all solid wall segments 24 inches (610 mm) or greater in length acted as one continuous solid wall length to resist a concentrated lateral force at the top of the wall. Because several separate wall segments of various lengths do not have as much capacity as one continuous solid wall segment to resist lateral loads, the results generated using the shear model were deemed unconservative for conditions where large amounts of openings may be possible (i.e., the building acts more like a concrete frame than a bearing wall system).

A flexure model assumed all solid wall segments 24 inches (610 mm) or greater in length acted separately and were equal in length; thereby uniformly distributing the load to each solid wall segment. Each solid wall segment was assumed to act as an unreinforced fixed-end cantilever, 8 feet (2.4 m) in height, to resist a concentrated lateral load at the top of the wall. Wall openings in residential structures are typically neither 8 feet (2.4 m) in height nor do they create solid wall segments completely unsupported at the top; therefore, several flexure models were investigated for wall segments with solid wall lengths ranging from 2 feet (610 mm) to 12 feet (3.7 m). All solid wall length segments occurring in a structure are rarely equal in length; however, the flexure models provided a range of acceptable values that were more conservative than the shear model discussed previously.

The solid wall percent values from the 6-, 8-, 9-, 10-, and 12-foot (1.8-, 2.4-, 2.7-, 3-, and 3.7-m) flexure models were used based on the values from the shear model. For example, as the shear model predicted lesser amounts of solid wall in lower load conditions, a progressively shorter segment flexure model was used. This resulted in a more uniform and logical set of bracing provisions that were more conservative and rational than that derived by the shear model alone. However, judgement was required to achieve the final values. The values were rounded to the nearest 5 percent and are listed in Table 5.3 of the *Prescriptive Method* as base percentages of solid wall length. Additional testing and research is needed in this area to determine more accurate solid wall length percentages when concrete walls contain perforations (i.e., wall openings) of various sizes.

While solid wall lengths less than 24 inches (610 mm) have historically been used in ICF residential construction, these segments are designed with knowledge of the specific load path to the wall segment and may require larger amounts of steel reinforcement. The 24-inch- (610-mm-) minimum solid wall length required at maximum intervals of 18 feet (5.5 m) on center and at all building corners is based on suggestions made by the steering committee, indeterminate wall opening placement locations and sizes, and similar requirements for masonry construction. The required percentage of solid wall length allows parallel shear to be resisted yet provides sufficient area for transmission of axial loads to the foundation. It is the intent that future testing will provide a more economical base percentage of solid wall length. The adjustment factors in Table 5.1 of the *Prescriptive Method* and the guidelines in Table 5.2 of the *Prescriptive Method* used to generate the minimum allowable percentage of solid wall length are based on the effects of building geometry on the wind or seismic loads received along a particular axis (i.e., wall line) of a building.

C5.2 Reinforcement around Openings

The requirements for number and placement of reinforcement around openings in Table 5.4 of the *Prescriptive Method* are based on ACI [C1], SBC [C7], and NBC [C5]. Per ACI [C1], the designer is required to provide two No. 5 bars around all window and door openings; this is considered excessive for residential loads. The SBC [C7] and NBC [C5] have clauses modifying this requirement to one No. 4 bar, provided that the vertical bars span continuously from support to support and that horizontal bars extend a minimum of 24 inches (610 mm) beyond the opening. The requirement for two No. 4 bars in conditions with wind speeds greater than 90 mph (145 km/hr) is provided to resist uplift loads.

C5.3 Lintels

C5.3.1 Load-Bearing ICF Wall Lintels

Lintels are horizontal members used to transfer wall, floor, roof, and attic dead and live loads around openings in load-bearing walls. Lintels are divided into three categories as follows:

- lintels in a one-story building or in the second story of a two-story building (supporting light-frame roof only);
- lintels in the first story of a two-story building where the second story is light-frame construction (supporting light-frame second story and roof); and
- lintels in the first story of a two-story building where the second story is ICF construction (supporting ICF second story and light-frame roof).

The following design assumptions were made in analyzing the lintels:

- Lintels are simply supported beams.
- A vertical core occurs at each end of the lintel for proper bearing.
- Lateral resistance is provided for the lintel by the floor or roof system above.
- Roof slopes range from 0:12 to 12:12.
- Deflection criterion is the clear span of the lintel, in inches, divided by 240.
- Ceilings, roofs, attics, and floors span the full width of the house (assume no interior load-bearing walls or beams).

- Floor and roof clear span is maximum 32 feet (9.8 m).
- Roof snow loads were calculated by multiplying the ground snow load by 0.7. Therefore, the roof snow load was taken as $P = 0.7P_g$, where P_g is the ground snow load in pounds per square foot.
- Loads experienced by the lintel are uniform loads and do not take into account any arching action that might occur because opening locations above the lintel cannot be determined for all cases.
- No. 3 stirrups are provided for spans greater than 4 feet (1.2 m) based on lintel test results; refer to *Lintel Testing for Reduced Shear Reinforcement in Insulating Concrete Form Systems* [C13].

All live and dead loads from the roof, attic, floor, wall above, and lintel itself were taken into account in the calculations using the ACI 318 [C1] load combination, $U = 1.4D + 1.7L$. This analysis is very conservative because the maximum snow, floor, and attic live loads have an extremely low probability of acting simultaneously throughout the life of the structure. In addition, the loads experienced by the lintel are based on the maximum allowable roof and floor clear span of 32 feet (9.8 m). Typically, the full dead load and a percentage of the live load is considered in lintel analysis where information regarding opening placement in the story is known The area of load combinations or lintels, particularly when multiple transient live loads from various areas of the building are considered, must be refined to produce more economical and rational designs.

The calculations are based on the lintel occurring in an above-grade wall with a floor live load of 30 psf (1.4 kPa). Due to the conservative nature of the lintel load analysis, the tables may be used for lintels located in foundation walls where the maximum floor live load is 40 psf (1.9 kPa) and additional wall dead loads from the story above are present.

Deflection limits are established primarily with regard to serviceability concerns. The intent is to prevent excessive deflection that may result in cracking of finishes. Windows and doors are especially sensitive to damage caused by lintel deflection; therefore, a conservative deflection limit of $L/480$ for service dead loads and sustained live loads is often suggested. This limit is very conservative when the installation of the window and door components is properly detailed. Accounting for the conservative lintel load analysis discussed above, $L/240$ for full service dead and live loads was used. The lintel section is assumed cracked and a stiffness factor of $0.1E_cI_g$ is used in accordance with test results and recommendations made in *Design Criteria for Insulating Concrete Form Wall Systems* [C11]. The reader is referred to Appendix B, *Engineering Technical Substantiation* for an example calculation of a load-bearing lintel.

Because the maximum allowable lintel spans seldom account for garage door openings in homes with a story above using a single No. 4 or No. 5 bottom bar for lintel reinforcement, Table 5.9 in the *Prescriptive Method* is provided for larger wall openings such as those commonly used for one- and two-car garage doors.

C5.3.2 ICF Gable End Wall Lintels (Non Load-Bearing)

Lintels are horizontal members used to transfer wall dead loads around openings in non load-bearing walls. Lintels are divided into two categories as follows:

- lintels in a one-story building or the second story of a two-story building and where the gable end wall is light-frame construction (supporting light-frame gable end wall); and
- lintels in the first story of a two-story building where the second story is ICF construction (supporting ICF second-story gable end wall).

The following design assumptions were made in analyzing the lintels:

- Lintels are simply supported beams.
- A vertical core occurs at each end of the lintel for proper bearing.
- Lateral resistance is provided for the lintel by the floor or roof system above.
- Deflection criterion is the clear span of the lintel, in inches, divided by 240.
- Lintels support only dead loads from the wall above and dead and live loads from a minimal tributary area of the roof.
- Loads experienced by the lintel are uniform loads and do not take into account any arching action that might occur above the lintel within a height equal to the lintel clear span because opening locations above the lintel cannot be determined for all cases.
- No. 3 stirrups are provided for spans greater than 4 feet (1.2 m) based on lintel test results; refer to *Lintel Testing for Reduced Shear Reinforcement in Insulating Concrete Form Systems* [C13].

Lintel dead weight and the dead load of the wall above were taken into account in the calculations using ACI 318 [C1] load combination, $U = 1.4D + 1.7L$. This analysis is conservative because arching action is not accounted for above the lintel within a height equal to the lintel clear span because wall opening locations above the lintel cannot be determined for all cases. The calculations are based on the lintel occurring in an above-grade wall. Due to the conservative nature of the lintel load analysis, the tables may be used for foundation walls where additional wall dead loads from the story above may be present.

Deflection limits are established primarily with regard to serviceability concerns. The intent is to prevent excessive deflection that may result in cracking of finishes. Windows and doors are especially sensitive to damage caused by lintel deflection; therefore, a conservative deflection limit of $L/480$ for service dead loads and sustained live loads is often suggested. This limit is very conservative when the installation of window and door components is properly detailed. Accounting for the conservative lintel load analysis discussed above, $L/240$ for full service dead and full service live loads was used.

The lintel section is assumed cracked and a stiffness factor of $0.1E_cI_g$ is used in accordance with test results and recommendations made in *Design Criteria for ICF Wall Systems* [C11]. The reader is referred to Appendix B, *Engineering Technical Substantiation* for an example calculation of a non load-bearing lintel.

C6.0 ICF CONNECTION REQUIREMENTS

C6.1 ICF Foundation Wall-to-Footing Connection

The requirements of the *Prescriptive Method* are based on typical residential construction practice for light-frame construction. Due to the heavier axial loads of ICF construction, frictional resistance at the footing-ICF wall interface is higher and provides a greater factor of safety than in light-frame residential construction.

C6.2 ICF Wall-to-Floor Connection

C6.2.1 Floor on ICF Wall Connection (Top-Bearing Connection)

The requirements of the *Prescriptive Method* are based on typical residential construction and CABO [C4] for foundations constructed of concrete masonry units.

C6.2.2 Floor Ledger-ICF Wall Connection (Side-Bearing Connection)

The requirements of the *Prescriptive Method* are based on the *Structural Design of Insulating Concrete Form Walls in Residential Construction* [C2]. Although other materials, such as cold-formed metal framing and concrete plank systems, may be used for the construction of floors in ICF construction, the majority of current ICF residential construction uses wood floor framing. Consult the manufacturer for proper connection details when using floor systems constructed of other materials. Consult a design professional when constructing buildings with floor systems which exceed the limits set forth in Table 1.1 of the *Prescriptive Method.*

C6.3 ICF Wall-to-Roof Connection

The roof framing-wood sill plate anchor requirements of the *Prescriptive Method* are based on typical residential construction and CABO [C4] for walls constructed of masonry.

The values listed in Table 6.3 for roof strap uplift requirements are based on wind uplift loads on the roof structure calculated in accordance with SBC [C7] taking into account the full dead load of the roof.

C7.0 UTILITIES

C7.1 Plumbing Systems

Due to the different ICF materials available, the reader is advised to refer to the local building code for guidance.

Typical construction practice with ICFs made of rigid plastic foam calls for cutting a chase into the foam for small pipes. Almost all ICFs made of rigid plastic foam will accommodate up to a 1-inch- (25-mm-) diameter pipe and some may accommodate up to a 2-inch- (51-mm-) diameter pipe. The pipes are typically fastened to the concrete with plastic or metal ties or concrete nails. The foam is then replaced with adhesive foam installed over the pipe. Larger pipes are typically installed on the inside face of the wall with a chase constructed around the pipe to conceal it; alternatively, pipes are rerouted through interior light-frame walls.

C7.2 HVAC Systems

Due to the different ICF materials available, the reader is advised to refer to the local building code for guidance.

ICF walls are considered to have high R-values and low air infiltration rates; however, due to the lack of test data, there is much debate on whether HVAC equipment can be sized smaller than in light-frame construction. Additional testing and monitoring are required.

C7.3 Electrical Systems

Due to the different ICF materials available, the reader is advised to refer to the local building code and the ICF manufacturer for guidance.

C8.0 CONSTRUCTION AND THERMAL GUIDELINES

The construction and thermal guidelines are provided to supplement the requirements of the *Prescriptive Method* and are considered good construction practices. These guidelines should not be considered comprehensive. Manufacturer's catalogs, recommendations, and other technical literature should also be consulted.

Proper fasteners and tools are essential to any trade. Tables C8.1 and C8.2 provide a list of fasteners and tools that are commonly used in residential ICF construction. Adhesives used on foam forms shall be compatible with the form material.

TABLE C8.1
TYPICAL FASTENERS FOR USE WITH ICFs

Fastener Type	Use/Application
Galvanized nails, ringed nails, and drywall screws	Attaching items to furring strips or form fastening surfaces
Adhesives	Attaching items to form for light- and medium-duty connections such as gypsum wallboard and base trim
J-bolt or steel straps	Attaching structural items to concrete core for heavy-duty connections such as floor ledger board and sill plate
Duplex nails	Attaching items to concrete core for medium-duty connections
Concrete nails or screw anchors	Attaching items to concrete core for medium-duty connections such as interior lumber partitions to exterior ICF walls

TABLE C8.2
RECOMMENDED TOOLS FOR ICF CONSTRUCTION

Tool	Use/ Application	Applicable Form Material
CUTTING		
Drywall saw	Small, straight, or curved cuts and holes	Foam
Keyhole saw	Precise holes for utility penetrations	All
PVC or miter saw	Small, straight cuts and for shaving edges of forms	Foam
Rasp or coarse sandpaper	Shaving edges of forms; removing small high spots after concrete pour	Foam
Hand saw	Fast, straight cuts	All
Circular saw	Fast, precise cuts; ensure proper blade is used	All
Reciprocating saw	Fast cuts, good for utility cuts, ensure proper blade is used	All
Thermal cutter	Fast, very precise cuts; removing large bulges in wall after concrete pour	Foam
Utility knife	Small, straight, or curved cuts and holes	Foam
Router	Fast, precise utility cuts; use with 1/2-inch drive for deep cutting	Foam
Hot knife	Fast, very precise utility cuts	Foam
MISCELLANEOUS		
Mason's trowel	Leveling concrete after pour; striking excess concrete from form after pour	All
	Applying thin mortar bed to forms	Composite
Wood glue, construction adhesive, or adhesive foam	Gluing forms together at joints	Foam
Cutter-bender	Cutting and bending steel reinforcement to required lengths and shapes	All
Small-gauge wire or precut tie wire or wire spool	Tying horizontal and vertical reinforcement together	All
Nylon tape	Reinforcing seams before concrete is poured	Foam
Nylon twine	Tying horizontal and vertical reinforcement together	All
Chalk line	Plumbing walls and foundation	All
Tin snips	Cutting metal form ties	Foam
MOVING/PLACING		
Forklift, manual lift, or boom or crane truck	Carrying large units or crates of units and setting them in place	All
Chute	Placing concrete in forms for below-grade pours	All
Line pump	Placing concrete in forms; use with a 2-inch hose	All
Boom pump	Placing concrete in forms; use with two "S" couplings and reduce the hose to a 2-inch diameter	All

C9.0 REFERENCES

[C1] *Building Code Requirements for Structural Concrete* (ACI 318-95). American Concrete Institute, Detroit, Michigan. 1995.

[C2] *Structural Design of Insulating Concrete Form Walls in Residential Construction.* Portland Cement Association, Skokie, Illinois. 1998.

[C3] *Minimum Design Loads for Buildings and Other Structures (*ASCE 7-93). American Society of Civil Engineers, New York, New York. 1993.

[C4] *One- and Two-Family Dwelling Code, 1995 Edition.* The Council of American Building Officials (CABO), Falls Church, Virginia. 1995.

[C5] *National Building Code.* Building Officials & Code Administrators International, Inc. (BOCA), Country Club Hills, Illinois. 1993.

[C6] *Uniform Building Code.* International Conference of Building Officials (ICBO), Whittier, California. 1994.

[C7] *Standard Building Code.* Southern Building Code Congress International, Inc. (SBCCI), Birmingham, Alabama. 1994.

[C8] *Guide to Residential Cast-in-Place Concrete Construction* (ACI 322R-84). American Concrete Institute, Detroit, Michigan. 1984.

[C9] ASTM C 31/C 31M-96 *Standard Practice for Making and Curing Concrete Test Specimens in the Field.* American Society for Testing and Materials (ASTM), West Conshohocken, Pennsylvania. 1997.

[C10] ASTM C 39-96 *Standard Test Method for Compressive Strength of Cylindrical Concrete Specimens.* American Society for Testing and Materials (ASTM), West Conshohocken, Pennsylvania. 1996.

[C11] *Design Criteria for Insulating Concrete Form Wall Systems, (RP 116),* Prepared for the Portland Cement Association by Construction Technology Laboratories, Inc., Skokie, Illinois. 1996.

[C12] *Mitigation of Moisture in Insulating Concrete Form Wall Systems,* Prepared for the Portland Cement Association by Construction Technology Laboratories, Inc., Skokie, Illinois. 1998.

[C13] *Lintel Testing for Reduced Shear Reinforcement in Insulating Concrete Form Systems,* Prepared for the U.S. Department of Housing and Urban Development, Portland Cement Association, and the National Association of Home Builders by NAHB Research Center, Inc., Upper Marlboro, Maryland. 1998.

APPENDIX A

ILLUSTRATIVE EXAMPLE

Purpose

Appendix A contains design examples illustrating the proper application of the different standards and specifications in the *Prescriptive Method.* It provides a step-by-step procedure on how to apply the requirements of the *Prescriptive Method* when designing a home. A typical residential building is used to demonstrate the application of insulating concrete form construction requirements.

Information is presented in both U.S. customary units and International System (SI) units except for reinforcement bar sizes which are only presented in U.S. customary units. Refer to Appendix C for the corresponding reinforcement bar size in SI units.

Building Design Criteria

The example building has the following characteristics:

Building type:	Two-story house (above an unfinished basement) with a center load-bearing beam supporting the first floor and a center load-bearing wall supporting the second floor
Building width:	28 feet (8.5 m)
Building length:	40 feet (12 m)
Maximum unbalanced backfill height	6 feet (1.8 m)
Soil-bearing capacity:	2,500 psf (120 kPa)
Basement wall height:	8 feet (2.4 m)
First story wall height:	9 feet (2.7 m)
Second story wall height:	8 feet (2.4 m)
Basement wall ICF type:	6-inch- (152-mm-) thick flat ICF concrete wall
First-story wall ICF type:	4.5-inch- (114-mm-) thick flat ICF concrete wall
Second-story wall ICF type:	4.5-inch- (114-mm-) thick flat ICF concrete wall
Floor joists:	Wood joists spaced at 24 inches (610 mm) on center
Roof framing:	Wood trusses spaced at 24 inches (610 mm) on center
Roof slope:	8:12

The following design criteria are applicable to the example home:

Ground Snow Load psf (kPa)	Wind Speed (mph) mph (km/hr)	Seismic Condition by Zone	First-Floor Live Load psf (kPa)	Second-Floor Live Load psf (kPa)	Floor Dead Load psf (kPa)	Equivalent Fluid Density pcf (kg/m^3)	Soil-Bearing Capacity psf (kPa)
45 (2.2)	80 (129)	2	40 (1.9)	30 (1.4)	10 (0.5)	30 (481)	2,500 (120)

Building elevations are shown in Figure A1.1 on the next page.

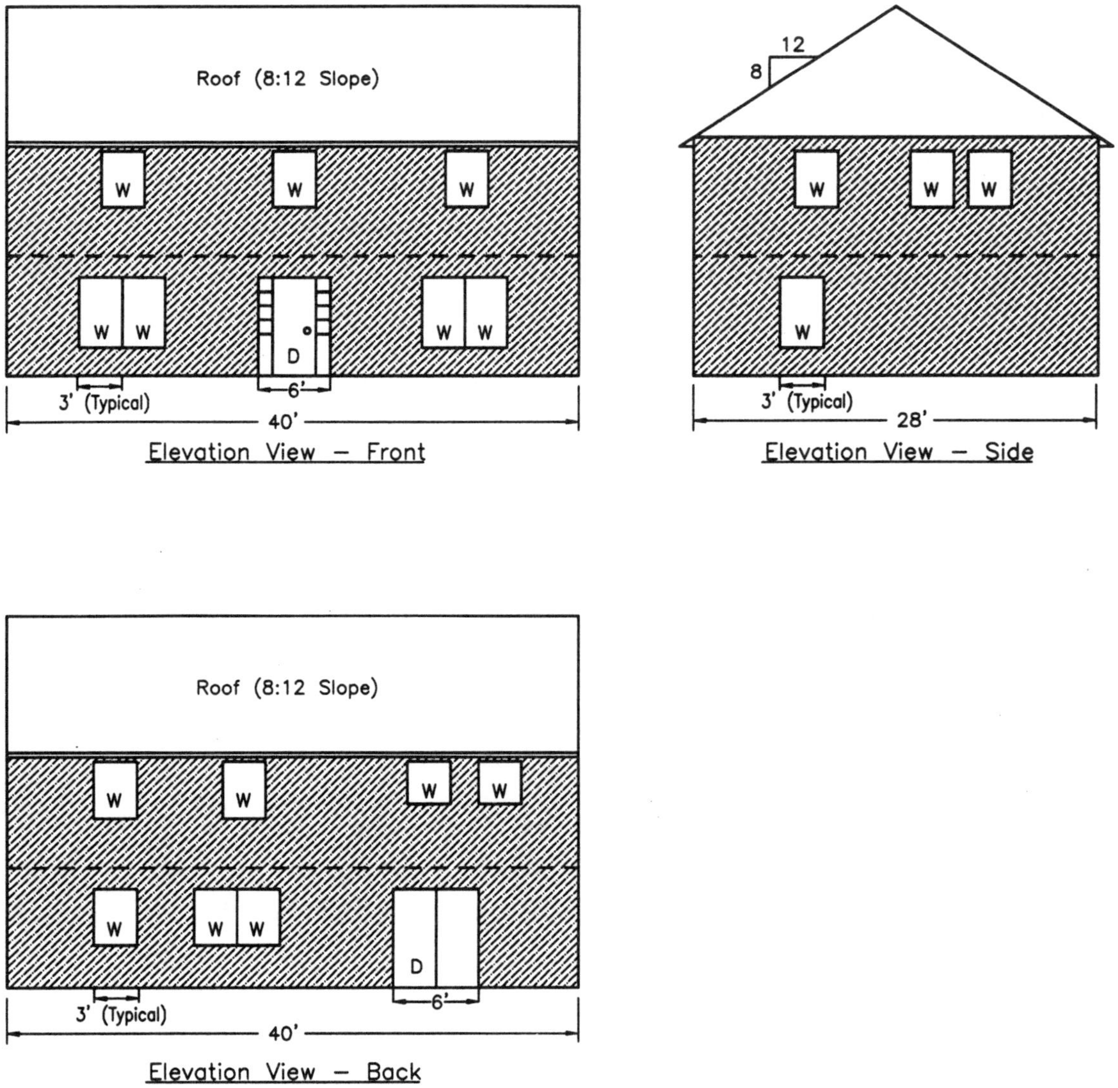

Figure A1.1 Building Elevations

Building Design Requirements Summary

The list below summarizes the requirements that result from applying the *Prescriptive Method* to the given building. A detailed description of the process is given in the following sections. Connection requirements are not highlighted in this Appendix since adequate details and tables are provided in the *Prescriptive Method.* The omission of detailed connection requirements from this Appendix is not, however, intended to diminish the importance of proper anchoring. The user should devote substantial efforts to connection requirements, which cannot be adequately conveyed in one building example.

TABLE A1.1
SUMMARY OF ILLUSTRATIVE EXAMPLE CALCULATIONS

Framing Description	Type	Thickness/ Size	Vertical Reinforcement Requirement	Horizontal Reinforcement Requirement	Reference
Footings	Concrete	23 inches by 7.6 inches	N/A	N/A	Section 2.0 Section 3.0 Table 3.1
Basement Walls	Flat ICF	6 inches thick	#3@12", #4@22", #5@30", or #6@40"	One No. 4 bar within 12 inches of the top of the wall story and one No. 4 bar near mid-height of the wall story	Section 2.0 Section 3.0 Table 3.3 Table 3.4
First-Story Walls	Flat ICF	4.5 inches thick	#3@14", #4@18", or #5@20"	One No. 4 bar within 12 inches of the top of the wall story and one No. 4 bar near third points of the wall story	Section 2.0 Section 4.0 Table 4.1 Table 4.2
Second-Story Walls	Flat ICF	4.5 inches thick	#4@32"	One No. 4 bar within 12 inches of the top of the wall story and one No. 4 bar near third points of the wall story	Section 2.0 Section 4.0 Table 4.1 Table 4.2
First-Story Solid Wall Length			10'-0" for 40-foot walls and 10'-4" for 28-foot walls		Section 5.0 Table 5.1 Table 5.2 Table 5.3
Second-Story Solid Wall Length			8'-0" for 40-foot walls and 5'-7" for 28-foot walls		Section 5.0 Table 5.1 Table 5.2 Table 5.3
First-Story Lintels	Flat lintels	4.5 inches thick by 22.4 inches deep	No. 4 bottom bar with No. 3 stirrups	One No. 4 bar within 12 inches of each side of the opening for the full height of the wall	Section 5.0 Table 5.4 Table 5.5

FOR SI: 1 foot = 0.3048 m; 1 inch = 25.4 mm

Footings

The design house is 28 feet (8.5 m) wide, the soil-bearing capacity is 2,500 psf (120 kPa), and the thickness of the basement wall is 6 inches (152 mm). The basement is supporting two ICF stories above. The minimum footing size for these conditions is established from Table 3.1 of the *Prescriptive Method* as 23 inches (584 mm) wide by 7.6 inches (193 mm) thick. The footings require a minimum concrete compressive strength of 2,500 psi (17 MPa).

TABLE 3.1
MINIMUM WIDTH OF ICF AND CONCRETE FOOTINGS FOR ICF WALLS[1,2]
(excerpt from the *Prescriptive Method*)

Maximum Number of Stories	Minimum Footing Width (inches)				
	Minimum Load-Bearing Value of Soil (psf)				
	2,000	2,500	3,000	3,500	4,000
5.5-Inch Flat, 6-Inch Waffle-Grid, or 6-Inch Screen-Grid ICF Wall Thickness[3]					
One Story	11	9	8	6	6
Two Story	20	15	13	11	10
Three Story[4]	31	**23**	18	15	13

ICF Basement Walls

The building is 28 feet (8.5 m) wide and is subject to lateral soil pressure with an equivalent fluid density of 30 pcf (481 kg/m^3). It has a maximum unbalanced backfill height of 6 feet (1.8 m), and its basement wall is 8 feet (2.4 m) in height. The minimum horizontal wall reinforcement in accordance with Table 3.3 of the *Prescriptive Method* is one No. 4 bar within 12 inches (305 mm) of the top of the wall story and one No. 4 bar near mid-height of the wall story. The ICF basement walls are assumed to be laterally supported at the top of the wall by the floor joists as required in Section 3.0 of the *Prescriptive Method.* In addition, using Table 3.4 of the *Prescriptive Method,* we find that the basement walls require minimum vertical wall reinforcement of one No. 3 bar, one No. 4 bar, one No. 5 bar, or one No. 6 bar spaced at 12 inches (305 mm), 22 inches (559 mm), 30 inches (762 mm), or 40 inches (1 m) on center respectively.

The basement walls require a minimum compressive strength of concrete of 2,500 psi (17.2 MPa) and a minimum steel tensile strength of 40 ksi (300 MPa) in accordance with Section 2.0 of the *Prescriptive Method.*

TABLE 3.3
MINIMUM HORIZONTAL WALL REINFORCEMENT FOR ICF BASEMENT WALLS
(excerpt from the *Prescriptive Method*)

Maximum Height of Basement Wall feet (meters)	Location of Horizontal Reinforcement
8 (2.4)	**One No. 4 bar within 12 inches (305 mm) of the top of the wall story and one No. 4 bar near mid-height of the wall story**
9 (2.7)	One No. 4 bar within 12 inches (305 mm) of the top of the wall story and one No. 4 bar near third points in the wall story

TABLE 3.4
MINIMUM VERTICAL WALL REINFORCEMENT FOR 5.5-INCH- (140-MM-)THICK FLAT ICF BASEMENT WALLS [1,2,3]
(excerpt from the *Prescriptive Method*)

Max. Height of Basement Wall (feet)	Maximum Unbalanced Backfill Height[4] (feet)	Minimum Vertical Reinforcement		
		Maximum Equivalent Fluid Density 30 pcf	Maximum Equivalent Fluid Density 45 pcf	Maximum Equivalent Fluid Density 60 pcf
8	4	#4@48"	#4@48"	#4@48"
	5	#4@48"	#3@12"; #4@22"; #5@32"; #6@40"	#3@8"; #4@14"; #5@20"; #6@26"
	6	**#3@12"; #4@22"; #5@30"; #6@40"**	#3@8"; #4@14"; #5@20"; #6@24"	#3@6"; #4@10"; #5@14"; #6@20"

ICF First-Story Walls

The building is 28 feet (8.5 m) wide and is subject to a wind speed of 80 mph (129 km/hr) and a ground snow load of 45 psf (2.2 kPa). The first-story walls are 9 feet (2.7 m) in height. In accordance with Table 4.1 of the *Prescriptive Method*, first-story walls require a minimum of one No. 4 horizontal bar within 12 inches (305 mm) of the top of the wall story and one No. 4 bar placed near third points in the wall story. Using Table 4.2 of the *Prescriptive Method*, we find that 4.5-inch- (114-mm-) thick first-story flat ICF walls supporting an ICF second story and light-frame roof also require a minimum of one vertical No. 3, one No. 4, or one No.5 bar spaced at 14 inches (356 mm), 18 inches (457 mm), or 20 inches (508 mm) respectively on center.

The first-story walls also require a minimum concrete compressive strength of 2,500 psi (17 MPa) and a minimum steel tensile strength of 40 ksi (300 MPa) in accordance with Section 2.0 of the *Prescriptive Method.*

TABLE 4.1
MINIMUM HORIZONTAL WALL REINFORCEMENT FOR ICF ABOVE-GRADE WALLS
(excerpt from the *Prescriptive Method*)

ICF Wall Type and Minimum Wall Thickness inches (mm)[1]	Maximum Height of Wall Story feet (meters)	Location of Horizontal Reinforcement
Flat 3.5 (89)	8 (2.4)	One No. 4 bar within 12 inches (305 mm) of the top of the wall story and one No. 4 bar near third points in the wall story
	9 (2.7)	**One No. 4 bar within 12 inches (305 mm) of the top of the wall story and one No. 4 bar near third points in the wall story**

TABLE 4.2
MINIMUM VERTICAL WALL REINFORCEMENT FOR FLAT ICF ABOVE-GRADE WALLS [1,2,3]
(excerpt from the *Prescriptive Method*)

Max. Wind Speed (mph)	Maximum Wall Height per Story (feet)	Minimum Vertical Reinforcement					
		Supporting Light-Frame Roof Only		Supporting Light-Frame Second Story and Roof		Supporting ICF Second Story and Light-Frame Roof	
		Minimum Wall Thickness (inches)					
		3.5	5.5	3.5	5.5	3.5	5.5
70	8	#4@32"	N/R	#4@32"	N/R	#4@32"	N/R
	9	#4@32"	N/R	#4@32"	N/R	#3@20"; #4@24"; #5@26"	N/R
	10	#4@32"	N/R	#4@32"	N/R	#3@20"; #4@24"; #5@26"	N/R
80	8	#4@32"	N/R	#4@32"	N/R	#4@32"	N/R
	9	#4@32"	N/R	#3@28"; #4@32"; #5@34"	N/R	**#3@14"; #4@18";#5@20"**	N/R

ICF Second-Story Walls

Applying the same sequence for the second-story walls as for the first-story walls, second-story walls require a minimum of one continuous No. 4 horizontal reinforcing bar placed within 12 inches (305 mm) of the top of the wall story and one No. 4 bar placed near third points in the wall story in accordance with Table 4.1 of the *Prescriptive Method.* Table 4.2 of the *Prescriptive Method* requires a minimum vertical wall reinforcement of one No. 4 bar spaced at 32 inches (813 mm) on center for flat walls with a thickness of 4.5 inches (114 mm) supporting a light-frame roof.

The second-story walls require a minimum concrete compressive strength of 2,500 psi (17.2 MPa) and a minimum steel tensile strength of 40 ksi (300 MPa) in accordance with Section 2.0 of the *Prescriptive Method.*

TABLE 4.1
MINIMUM HORIZONTAL WALL REINFORCEMENT FOR ICF ABOVE-GRADE WALLS
(excerpt from the *Prescriptive Method*)

ICF Wall Type and Minimum Wall Thickness inches (mm)[1]	Maximum Height of Wall Story feet (meters)	Location of Horizontal Reinforcement
Flat 3.5 (89)	8 (2.4)	**One No. 4 bar within 12 inches (305 mm) of the top of the wall story and one No. 4 bar near third points in the wall story**
	9 (2.7)	One No. 4 bar within 12 inches (305 mm) of the top of the wall story and one No. 4 bar near third points in the wall story

TABLE 4.2
MINIMUM VERTICAL WALL REINFORCEMENT FOR FLAT ICF ABOVE-GRADE WALLS [1,2,3]
(excerpt from the *Prescriptive Method*)

Max. Wind Speed (mph)	Maximum Wall Height per Story (feet)	Minimum Vertical Reinforcement					
		Supporting Light-Frame Roof Only		Supporting Light-Frame Second Story and Roof		Supporting ICF Second Story and Light-Frame Roof	
		Minimum Wall Thickness (inches)					
		3.5	5.5	3.5	5.5	3.5	5.5
70	8	#4@32"	N/R	#4@32"	N/R	#4@32"	N/R
	9	#4@32"	N/R	#4@32"	N/R	#3@20"; #4@24"; #5@26"	N/R
	10	#4@32"	N/R	#4@32"	N/R	#3@20"; #4@24"; #5@26"	N/R
80	8	**#4@32"**	N/R	#4@32"	N/R	#4@32"	N/R
	9	#4@32"	N/R	#3@28"; #4@32"; #5@34"	N/R	#3@14"; #4@18"; #5@20"	N/R

Minimum Length of ICF Wall without Openings

To determine the minimum percentage of solid wall length required for the example house, Section 5.0 and Tables 5.1 through 5.3 of the *Prescriptive Method* are used. The design house is a two-story home, 28'-0" (8.5 m) wide by 40 feet (12 m) long. It has a roof slope of 8:12 and is located in an 80 mph (129 km/hr) wind area and Seismic Zone 2. Each above-grade wall story is 4.5 inches (114 mm) thick.

The closest roof slope category in Table 5.3 of the *Prescriptive Method* is 9:12. The design roof slope is 8:12; therefore, the following calculations may be used to determine the exact solid wall length required.

FIRST-STORY WALLS

In accordance with Table 5.2 of the Prescriptive Method, the minimum solid wall length is the greater of the following values.

(a). The base percentage of solid wall length is based on wind speed in Table 5.3 multiplied by the appropriate factor from Table 5.1 of the *Prescriptive Method* for the 28-foot (8.5-m) and 40-foot (12-m) walls.

60 percent (9:12) minus 40 percent (6:12) equals three equal percentage increments of 6.67 percent; (60% - 40%)/3 = 6.67%. 60 percent (9:12) minus 6.67 percent (1:12) indicates the base percentage of solid wall length is *53 percent.*

The building aspect ratio is approximately 1.4 (40 feet/28 feet =1.4). Therefore, the base percentage of solid wall length may be adjusted as follows.

- The base percentage of solid wall length may be reduced by the appropriate adjustment factor from Table 5.1 of the *Prescriptive Method* to (0.7)(53) = 37 percent for the 28-foot (8.5-m) wall.
- The base percentage of solid wall length may be reduced by the appropriate adjustment factor from Table 5.1 of the *Prescriptive Method* to (0.4)(53) = 21 percent for the 40-foot (12-m) wall.

(b). The base percentage of solid wall length for Seismic Zone 2 in Table 5.3 of the *Prescriptive Method* is *35 percent.* This base percentage of solid wall length may not be adjusted for the 28-foot (8.5-m) wall; however, the base percentage of solid wall length may be divided by the building aspect ratio for the 40-foot (12-m) wall.

- 35 percent for the 28-foot (8.5-m) wall
- 35 percent/1.4 = 25 percent for the 40-foot (12-m) wall

(c). A minimum of 20 percent is required for walls supporting an ICF or light-frame second story and a light-frame roof.

The greatest value determined in (a) through (c) above for the 28-foot (8.5-m) wall is 37 percent; thus, the 28-foot (8.5-m) walls require (0.37)(28) = 10'-4" (3.2 m) of solid wall length. The greatest value determined in (a) through (c) above for the 40-foot (12-m) wall is 25 percent; thus, the 40-foot (12-m) walls require (0.25)(40) = 10'-0" (3 m) of solid wall length. Remember that Section 5.0 requires a minimum length of 24 inches (610 mm) of solid wall segment, extending the full height of each wall story, at all corners of exterior walls and that 2-foot (610-mm) wall segments shall not exceed the maximum allowable spacing of 18 feet (5.5 m) on center. Both of these requirements are met; therefore, the calculated minimum allowable percentages of solid wall length may be used.

SECOND-STORY WALLS

In accordance with Table 5.2 of the *Prescriptive Method*, the minimum solid wall length is the greater of the following values.

(a). The base percentage of solid wall length is based on wind speed in Table 5.3 multiplied by the appropriate factor from Table 5.1 of the *Prescriptive Method* for the 28-foot (8.5-m) and the 40-foot (12-m) walls.

25 percent (9:12) minus 15 percent (6:12) equals three equal percentage increments of 3.33 percent; (25% - 15%)/3 = 3.33%. 25 percent (9:12) minus 3.33 percent (1:12) indicates the base percentage of solid wall length is *21.7 percent.*

The building aspect ratio is approximately 1.4 (40 feet/28 feet =1.4). Therefore, the base percentage of solid wall length may be adjusted as follows.

- The base percentage of solid wall length may be reduced by the appropriate adjustment factor from Table 5.1 of the *Prescriptive Method* to (0.7)(21.7) = <u>15 percent for the 28-foot (8.5-m) wall</u>.
- The base percentage of solid wall length may be reduced by the appropriate adjustment factor from Table 5.1 of the *Prescriptive Method* to (0.4)(21.7) = <u>9 percent for the 40-foot (12-m) wall</u>.

(b). The base percentage of solid wall length for Seismic Zone 2 in Table 5.3 of the *Prescriptive Method* is *20 percent.* This base percentage of solid wall length may not be adjusted for the 28-foot (8.5-m) wall; however, the base percentage of solid wall length may be divided by the building aspect ratio for the 40-foot (12-m) wall.

- <u>20 percent for the 28-foot (8.5-m) wall</u>
- 20 percent/1.4 = <u>14 percent for the 40-foot (12-m) wall</u>

(c). A minimum of <u>15 percent</u> is required for walls supporting a light-frame roof.

The greatest value determined in (a) through (c) above for the 28-foot (8.5-m) wall is 20 percent; thus, the 28-foot (8.5-m) walls require (0.20)(28) = 5'-7" (1.7 m) of solid wall length. The greatest value determined in (a) through (c) above for the 40-foot (12-m) wall is 15 percent; thus, the 40-foot (12-m) walls require (0.15)(40) = 6'-0" (1.8 m) of solid wall length. Remember that Section 5.0 requires a minimum length of 24 inches (610 mm) of solid wall segment, extending the full height of each wall story, at all corners of exterior walls and that 2-foot (610-mm) wall segments shall not exceed the maximum allowable spacing of 18 feet (5.5 m) on center. These requirements are not met using the calculated minimum allowable percentage of solid wall length for the 40-foot (12-m) wall; therefore, the 40-foot (12-m) wall requires a minimum of 8'-0" solid wall length to meet the maximum allowable spacing of 18 feet (5.5 m) on center.

TABLE 5.1
ADJUSTMENT FACTORS FOR USE WITH TABLE 5.2[1]
(excerpt from the *Prescriptive Method*)

Building Aspect Ratio[2]	Adjustment Factor	
	Parallel to the Shorter Building Dimension or Endwall, W	Parallel to the Longer Building Dimension or Sidewall, L
2.0	1.0	0.25
1.8	0.9	0.30
1.6	0.8	0.35
1.4	**0.7**	**0.40**

TABLE 5.2
MINIMUM PERCENTAGE OF SOLID WALL LENGTH ALONG EXTERIOR WALL LINES[1]
(excerpt from the *Prescriptive Method*)

Exterior Wall Line Category	Site Wind and Seismic Conditions[2,3]	
	All Wind Speeds and Seismic Zones 0 and 1	**All Wind Speeds and Seismic Zone 2**
Parallel to the Shorter Building Dimension or Endwall, W	The minimum percentage of solid wall length shall be the maximum of the following: • The base percentage of solid wall length based on wind speed in Table 5.3 multiplied by the appropriate factor from Table 5.1. • 15 percent for walls supporting a light-frame roof or 20 percent for walls supporting an ICF or light-frame second story and light-frame roof.	**The minimum percentage of solid wall length shall be the maximum of the following: • The base percentage of solid wall length based on wind speed in Table 5.3 multiplied by the appropriate factor from Table 5.1. • The base percentage of solid wall length for Seismic Zone 2 in Table 5.3 without adjustment. • 15 percent for walls supporting a light-frame roof or 20 percent for walls supporting an ICF or light-frame second story and light-frame roof.**
Parallel to the Longer Building Dimension or Sidewall, L	The minimum percentage of solid wall length shall be the maximum of the following: • The base percentage of solid wall length based on wind speed in Table 5.3 multiplied by the appropriate factor from Table 5.1. • 15 percent for walls supporting a light-frame roof or 20 percent for walls supporting an ICF or light-frame second story and light-frame roof.	**The minimum percentage of solid wall length shall be the maximum of the following: • The base percentage of solid wall length based on wind speed in Table 5.3 multiplied by the appropriate factor from Table 5.1. • The base percentage of solid wall length for Seismic Zone 2 in Table 5.3 divided by the building aspect ratio. • 15 percent for walls supporting a light-frame roof or 20 percent for walls supporting an ICF or light-frame second story and light-frame roof.**

[1] Refer to Figure 5.1 for the definition of L, W, and the building aspect ratio.
[2] The percent of solid wall in Seismic Zones 3 and 4, regardless of wind speed, shall be determined by an approved design.
[3] In no case shall the spacing of wall segments at least 24 inches (610 mm) in length exceed 18 feet (5.5 m) on center.

TABLE 5.3
BASE PERCENTAGE OF SOLID WALL LENGTH ALONG EXTERIOR WALL LINES[1,2,3]
(excerpt from the *Prescriptive Method*)

ICF Wall Type and Minimum Wall Thickness[4] (inches)	Max. Roof Slope	Base Solid Wall Length (percent)									
		Wall Supporting Light-Frame Roof					Wall Supporting ICF or Light-Frame Second Story and Light-Frame Roof				
		Maximum Wind Speed (mph)									
		70	80	90	100	110	70	80	90	100	110
Flat 3.5	3:12	15	15	15	15	20	30	35	40	50	55
	6:12	15	**15**	20	20	25	30	**40**	50	55	60
	9:12	20	**25**	30	40	45	45	**60**	70	85	95
	12:12	25	35	40	50	60	50	65	80	95	100
Flat 5.5	3:12	15	15	15	15	15	20	25	30	40	40
	6:12	15	15	15	15	20	20	30	35	40	45
	9:12	15	15	20	25	30	35	45	50	60	70
	12:12	20	20	25	35	40	35	50	55	70	75
Flat 7.5	3:12	15	15	15	15	15	20	20	25	30	30
	6:12	15	15	15	15	15	20	20	25	30	35
	9:12	15	15	15	20	25	25	30	40	45	50
	12:12	15	20	20	25	30	30	35	40	50	55
Waffle-Grid 6	3:12	15	15	15	15	20	25	30	35	45	50
	6:12	15	15	20	20	25	25	35	45	50	55
	9:12	20	20	25	35	40	40	55	60	75	85
	12:12	25	30	35	45	50	45	60	70	85	90
Waffle-Grid 8	3:12	15	15	15	15	15	20	25	30	35	35
	6:12	15	15	15	15	20	20	25	30	35	40
	9:12	15	15	20	25	30	30	40	45	55	60
	12:12	20	20	25	30	35	35	40	50	60	65
Screen-Grid 6	3:12	15	15	20	20	25	30	40	45	55	60
	6:12	15	20	25	30	35	30	40	50	60	70
	9:12	25	30	45	50	65	50	65	80	90	100
	12:12	35	40	55	65	80	55	70	85	100	100
Seismic Zone 2											
Flat, 3.5	N/A	**20**					**35**				
Flat, 5.5	N/A	15					30				
Flat, 7.5	N/A	15					25				
Waffle-Grid, 6	N/A	20					35				
Waffle-Grid, 8	N/A	20					30				
Screen-Grid, 6	N/A	25					45				

For SI: 1 inch = 25.4 mm; 1 mph = 1.6093 km/hr

1 Linear interpolation between roof slopes shall be permitted.
2 Base percentages are applicable for maximum 10-foot (3.0-m) wall story heights.
3 N/A indicates not applicable.
4 Actual thickness is shown for flat walls while nominal thickness is given for waffle- and screen-grid walls . Refer to Section 2.0 for actual waffle- and screen-grid thickness and dimensions.

Wall Openings and Lintels

In this example, there are two opening sizes: 6-foot- (1.8-m-) wide openings for the entrance doorway, slider door, and double windows in the first wall story and 3-foot- (0.9-m-) wide openings throughout the remainder of the building. Tables 5.4 and 5.5 of the *Prescriptive Method* are used in this exercise to determine the lintel type and reinforcement requirements. For the 3-foot- (0.9-m-) wide openings, Table 5.4 of the *Prescriptive Method* requires one No. 4 bar to be placed within 12 inches (305 mm) from .the bottom and top of the opening with a minimum embedment of 24 inches (610 mm) beyond each side of the opening. For the 6-foot- (1.8-m-) wide openings, Table 5.4 of the *Prescriptive Method* requires a lintel over the opening in accordance with Section 5.0.

The 6-foot- (1.8-m-) wide openings are located in the first-story load-bearing walls of the building. First-story walls were previously stated as 4.5-inch- (114-mm-) thick flat ICF walls; therefore, Table 5.5 of the *Prescriptive Method* is used to determine the lintel's reinforcement requirements. The building is 28 feet (8.5 m) wide and subject to a ground snow load of 45 psf (2.2 kPa). Since the lintels are located in the first-story wall, they support one ICF story and a light-frame roof and ceiling. Table 5.5 of the *Prescriptive Method* illustrates that a 4.5-inch (114-mm) lintel thickness and 24-inch (610-mm) lintel depth will suffice since the lintel's maximum allowed span is 6'-1" (1.9 m). Footnote 2 of Table 5.5 of the *Prescriptive Method* allows the user to interpolate between snow loads and between lintel depths. A maximum lintel span of 6'-6" (2m) and a minimum lintel depth of 22.4 inches (569 mm) result when interpolation is used for a 45 psf (2.2 kPa) ground snow load.

In addition, Table 5.4 of the *Prescriptive Method* requires a minimum vertical reinforcement of one No. 4 bar within 12 inches (305 mm) of each side of the 6-foot- (1.8-m-) wide opening for the full height of the wall in conditions where wind speeds are 80 mph (129 km/hr).

TABLE 5.4
MINIMUM WALL OPENING REINFORCEMENT REQUIREMENTS IN ICF WALLS
(excerpt from the *Prescriptive Method*)

Wall Type and Opening Width, L feet (m)	Minimum Horizontal Opening Reinforcement	Minimum Vertical Opening Reinforcement
Flat, Waffle-, and Screen-Grid: L < 2 (0.61)	None Required	None Required
Screen-Grid: 2 (0.61) ≤ L ≤ 4 (1.2)	One No. 4 bar a minimum of 1.5 inches (38 mm) and a maximum of 2.5 inches (64 mm) from the top of the opening. One No. 4 bar within 12 inches (305 mm) of the bottom of the opening. Each No. 4 bar shall extend a minimum of 24 inches (610 mm) beyond the limits of the opening.	None Required
Flat and Waffle-Grid: 2 (0.61) ≤ L < 4 (1.2)	**One No. 4 bar within 12 inches (305 mm) of the top of the opening. One No. 4 bar within 12 inches (305 mm) from the bottom of the opening. Each No. 4 bar shall extend a minimum of 24 inches (610 mm) beyond the limits of the opening.**	**None Required**
Flat, Waffle-, and Screen-Grid: L ≥ 4 (1.2)	**Provide lintels in accordance with Section 5.0. Top and bottom lintel reinforcement shall extend a minimum of 24 inches (610 mm) beyond the limits of the opening.**	**In conditions with wind speeds less than or equal to 90 mph (145 km/hr), provide one No. 4 bar for the full height of the wall story within 12 inches (305 mm) of each side of the opening.** **In conditions with wind speeds greater than 90 mph (145 km/hr), provide two No. 4 bars for the full height of the wall story within 12 inches (305 mm) of each side of the opening.**

TABLE 5.5
MAXIMUM ALLOWABLE CLEAR SPANS FOR
FLAT ICF WALL LINTELS IN LOAD-BEARING WALLS [1,2,3]
NO. 4 BOTTOM BAR SIZE
(excerpt from the *Prescriptive Method*)

Minimum Lintel Thickness, T (inches)	Minimum Lintel Depth, D (inches)	Maximum Clear Span					
		Supporting Light-Frame Roof Only		Supporting Light-Frame Second Story and Roof		Supporting ICF Second Story and Light-Frame Roof	
		Maximum Ground Snow Load (psf)					
		30	70	30	70	30	70
3.5	8	4'-9"	4'-2"	3'-10"	3'-4"	3'-5"	3'-1"
	12	6'-8"	5'-5"	5'-0"	4'-5"	4'-6"	4'-0"
	16	7'-11"	6'-5"	6'-0"	5'-3"	5'-4"	4'-10"
	20	8'11"	7'-4"	6'-9"	6'-0"	6'-1"	5'-6"
	24	9'-10"	8'-1"	7'-6"	6'-7"	6'-9"	**6'-1"**
5.5	8	5'-2"	4'-2"	3'-10"	3'-5"	3'-5"	3'-1"
	12	6'-8"	5'-5"	5'-0"	4'-5"	4'-6"	4'-1"
	16	7'-10"	6'-5"	6'-0"	5'-3"	5'-4"	4'-10"
	20	8'-10"	7'-3"	6'-9"	6'-0"	6'-1"	5'-6"
	24	9'-8"	8'-0"	7'-5"	6'-7"	6'-8"	6'-0"

ICF Foundation Wall-to-Footing Connection

Section 6.0 of the *Prescriptive Method* requires that no dowels be installed across the ICF wall-footing interface because the interior floor slab will be poured before backfilling and installed in accordance with Figure 3.3 of the *Prescriptive Method.*

ICF Basement Wall-to-Floor Connection

The design house is 28 feet (8.5 m) wide and has wood floor joists spaced at 24 inches (610 mm) on center. A wood ledger beam is used to support the floor joists; refer to Section 6.0. A 2-inch (51-mm) wood ledger with a 1/2-inch (13-mm) anchor bolt is chosen. Assuming that a load-bearing beam is placed beneath the floor at mid-span, the resulting floor clear span is 14 feet (4.3 m). Table 6.1 of the *Prescriptive Method* requires a minimum staggered 1/2-inch- (13-mm-) diameter anchor bolt spaced at 12 inches (305 mm) on center. Alternatively, two 1/2-inch- (13-mm-) diameter anchor bolts spaced at 24 inches (610 mm) on center, staggered 5/8-inch- (15.8-mm-) diameter anchor bolt spaced at 16 inches (406 mm) on center, or two 5/8-inch- (15.8-mm-) diameter anchor bolts spaced at 32 inches (813 mm) on center may also be used.

TABLE 6.1
FLOOR LEDGER-ICF WALL CONNECTION (SIDE-BEARING CONNECTION) REQUIREMENTS[1,2,3]
(excerpt from the *Prescriptive Method*)

Maximum Floor Clear Span[4] (feet)	Maximum Anchor Bolt Spacing[5] (inches)			
	Staggered 1/2-Inch-Diameter Anchor Bolts	Staggered 5/8-Inch-Diameter Anchor Bolts	Two 1/2-Inch-Diameter Anchor Bolts[6]	Two 5/8-Inch-Diameter Anchor Bolts[6]
8	18	20	36	40
10	16	18	32	36
12	14	18	28	36
14	**12**	**16**	**24**	**32**

ICF Wall-To-Roof Connection

Section 6.0, Table 6.2, and Figure 6.8 of the *Prescriptive Method* are used to determine the top of the ICF wall-to-roof connection requirements. It is assumed that the design house has wood trusses in lieu of wood rafters and is located in an 80 mph (129 km/hr) wind area. Table 6.2 of the *Prescriptive Method* requires a minimum of a 1/2-inch (13-mm) anchor bolt spaced at 6 feet (1.8 m) on center; no roof strap is required.

TABLE 6.2
TOP SILL PLATE-ICF WALL CONNECTION REQUIREMENTS
(excerpt from the *Prescriptive Method*)

Maximum Wind Speed (mph)	Maximum Anchor Bolt Spacing[1]	
	1/2-Inch-Diameter Anchor Bolt	5/8-Inch-Diameter Anchor Bolt
70	6'-0"	8'-0"
80	**6'-0"**	8'-0"
90	6'-0"	8'-0"
100	4'-0"	6'-0"
110	4'-0"	4'-0"

For SI: 1 foot = 0.3048 m; 1 inch = 25.4 mm; 1 mph = 16.09344 km/hr

[1] Minimum anchor bolt embedment length shall be 4 inches (102 mm).

APPENDIX B

ENGINEERING TECHNICAL SUBSTANTIATION

INTRODUCTION

The *Engineering Technical Substantiation* is provided as a supplemental information package to illustrate the intent and basis for the development of the *Prescriptive Method.* Structural calculations illustrate the method for calculating the reinforcement requirements for the walls and lintels; supplemental equations and reference standards substantiate the examples. The example calculations are not intended to be inclusive of all design considerations for a given application but rather are intended to illustrate the application of the requirements of the *Prescriptive Method.*

Information is presented in both U.S. customary units and International System (SI) units except for reinforcement bar sizes which are only presented in U.S. customary units. Refer to Appendix C for the corresponding reinforcement bar size in SI units.

1.0 GENERAL

1.1 Load Calculations

Roof Loads

Roof snow loads were calculated by using ASCE 7 [B1], multiplying the ground snow load by 0.7. Therefore, roof snow load was taken as $P = 0.7P_g$, where P_g is the ground snow load in pounds per square foot. Off-balance snow loads were not considered since heavier gravity loads generally improve the performance of concrete walls. Also, this approach has proven successful based on past performance and experience in residential construction when building and roof spans are relatively small.

Wind Loads

Wind loads were based on fastest-mile wind speeds ranging from 70 to 110 mph (113 to 177 km/hr). Wind pressures were calculated in accordance with SBC [B2] by using components and cladding coefficients for enclosed buildings, interior zone, and a mean roof height of 35 feet (10.7 m). The calculated wind load is the product of the velocity pressure, the pressure coefficient, and the use factor as shown in Figure B1.1. The component and cladding wind loads were used to determine out-of-plane bending and shear in the walls. Main wind force resisting system (MWFRS) loads were also determined assuming an enclosed building condition in accordance with SBC [B2]. The distribution of lateral wind loads to components or assemblies comprising the MWFRS was achieved by a standard tributary area method. Wind forces on all windward and leeward surfaces were considered when determining MWFRS loads.

Wind Speed mph (km/hr)	Calculation	Components and Cladding Wind Load psf (kPa)
70 (113)	12.8 x 1.25 x 1.0 =	16 (0.77)
80 (129)	16.7 x 1.25 x 1.0 =	21 (1.0)
90 (145)	21.1 x 1.25 x 1.0 =	27 (1.3)
100 (161)	26.0 x 1.25 x 1.0 =	32 (1.5)
110 (177)	31.5 x 1.25 x 1.0 =	40 (1.9)

Figure B1.1 Calculated Wind Loads

1.2 ICF Foundation Wall Design Assumptions

Four different construction cases were investigated to cover the range of residential construction possibilities. Descriptions of the four construction cases follow.

A1:

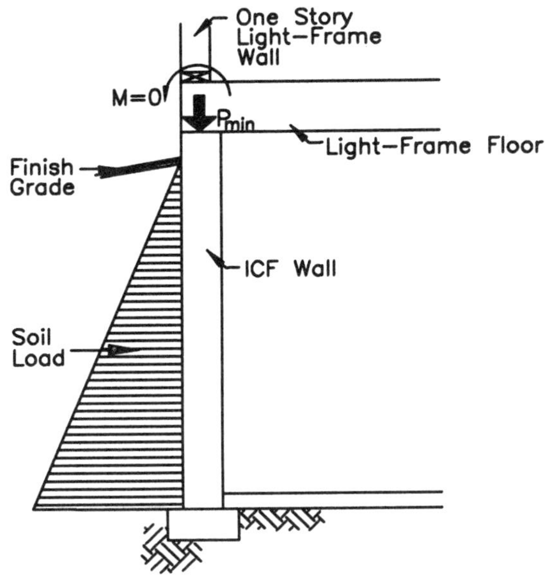

Supporting One-Story Light-Frame Construction

- Smallest practical axial load with light-frame construction due to dead and live loads
- No eccentricity; assume floor bearing on foundation wall
- Earth load varies from 30 pcf (481 kg/m^3) to 60 pcf (960 kg/m^3) by equivalent fluid density

A2:

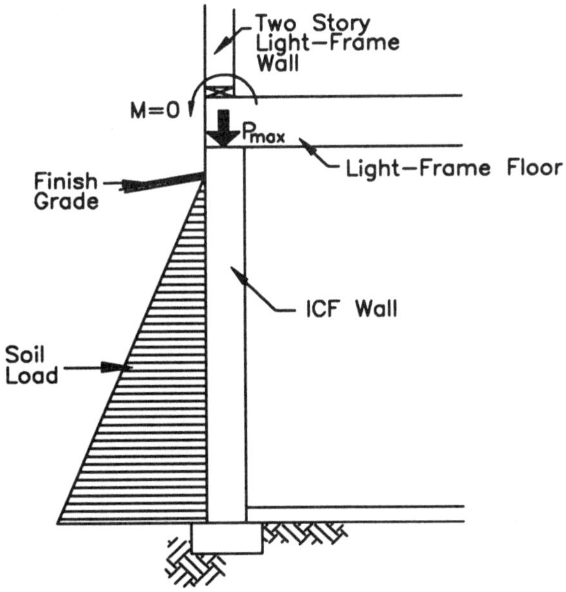

Supporting Two-Story Light-Frame Construction

- Largest axial design load with light-frame construction due to dead and live loads
- No eccentricity; assume floor bearing on foundation wall
- Earth load varies from 30 pcf (481 kg/m^3) to 60 pcf (960 kg/m^3) by equivalent fluid density

A3:

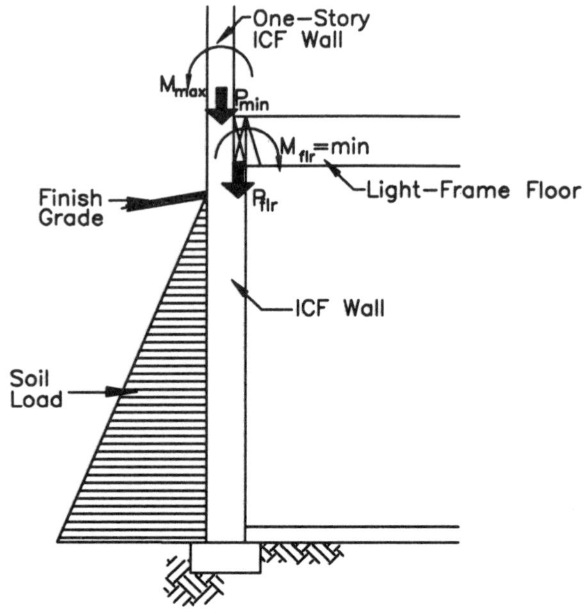

Supporting One-Story ICF Construction

- Smallest practical axial design load with ICF construction due to dead and live loads
- Smallest floor eccentricity; assume side-bearing ledger board for floor connection
- Largest practical wall eccentricity; assume ICF wall thickness for above-grade wall is smaller than ICF wall thickness for foundation wall
- Earth load varies from 30 pcf (481 kg/m^3) to 60 pcf (960 kg/m^3) by equivalent fluid density

A4:

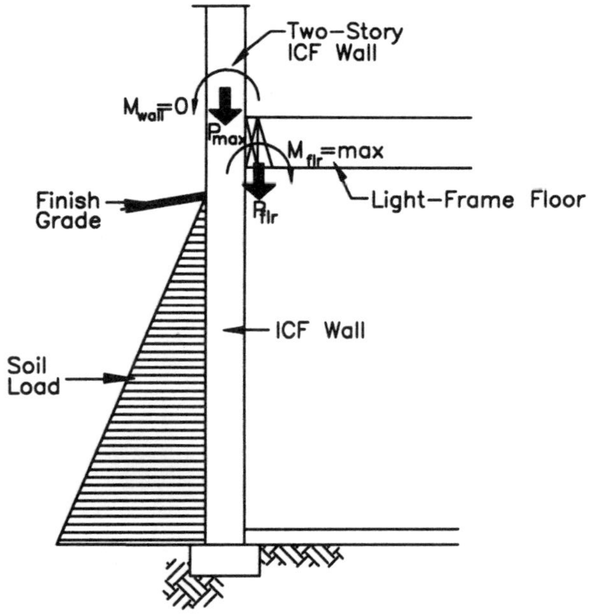

Supporting Two-Story ICF Construction

- Largest axial design load with ICF construction due to dead and live loads
- Largest floor eccentricity; assume sidc-bcaring ledger board for floor connection
- No wall eccentricity; assume ICF wall thickness for above-grade wall is equal to ICF wall thickness for foundation wall
- Earth load varies from 30 pcf (481 kg/m^3) to 60 pcf (960 kg/m^3) by equivalent fluid density

The following values were used to design the foundation walls; refer to Figure B1.2. The foundation wall height, unbalanced backfill height, and earth load vary. The vertical wall reinforcements listed in the minimum vertical wall reinforcement tables of the *Prescriptive Method* are the results from calculations based on the governing construction case (A1 through A4) and ACI [B3] Load Combination.

The first-floor live load induces a negative moment on the foundation wall in construction case A3; therefore, a sustained live load of 10 psf (0.48 kPa) was considered to produce the largest practical positive moment on the foundation wall resulting from backfill earth pressure, above-grade wall eccentricity, and minimum floor eccentricity.

Vinyl siding was considered as the exterior wall covering in all the construction cases for the above-grade walls. In hindsight, the wall loading for the above-grade walls should reflect a brick veneer for construction cases A2 and A4; however, construction case A1, with the lightest practical loads governed the minimum vertical wall reinforcement required for foundation walls.

BUILDING GEOMETRY	Units	FOUNDATION CONSTRUCTION CASE			
		A1	**A2**	**A3**	**A4**
Maximum Building Plan Dimension	feet (m)	60 (18)	60 (18)	60 (18)	60 (18)
Roof Slope	----	0:12	12:12	0:12	12:12
Roof Clear Span	feet (m)	40 (12)	40 (12)	40 (12)	40 (12)
First-Floor Clear Span	feet (m)	16 (4.9) to 32 (9.8)	32 (9.8)	16 (4.9) to 32 (9.8)	32 (9.8)
Second-Floor Clear Span	feet (m)	N/A	32 (9.8)	N/A	32 (9.8)
First-Story Wall Height	feet (m)	8 (2.4)	10 (3)	8 (2.4)	10 (3)
Second-Story Wall Height	feet (m)	N/A	10 (3)	N/A	10 (3)
First-Story Wall Thickness	inches (mm)	3.5 (89) light-frame	5.5 (140) light-frame	3.5 (89) flat; 5.0 (127) waffle; 5.5 (140) screen	7.5 (191) flat; 7.0 (178) waffle; 5.5 (140) screen
Second-Story Wall Thickness	inches (mm)	N/A	5.5 (140) light-frame	N/A	5.5 (140) flat; 5.0 (127) waffle; 5.5 (140) screen
Foundation Wall Thickness	inches (mm)	vary	vary	vary	vary
DEAD LOADS					
First Floor	psf (kPa)	10 (0.48)	15 (0.72)	10 (0.48)	15 (0.72)
Second Floor	psf (kPa)	N/A	15 (0.72)	N/A	15 (0.72)
Roof and Ceiling	psf (kPa)	12 (0.58)	16 (0.77)	12 (0.58)	16 (0.77)

BUILDING GEOMETRY	Units	FOUNDATION CONSTRUCTION CASE			
		A1	A2	A3	A4
First-Story Wall	psf (kPa)	6 (0.3)	11 (0.5)	50 (2.4) flat; 55 (2.6) waffle; 50 (2.4) screen	100 (4.8) flat; 75 (3.6) waffle; 50 (2.4) screen
Second-Story Wall	psf (kPa)	N/A	11 (0.5)	N/A	75 (3.6) flat; 55 (2.6) waffle; 50 (2.4) screen
Foundation Wall	psf (kPa)	vary	vary	vary	vary
LIVE LOADS					
First Floor	psf (kPa)	40 (1.9)	40 (1.9)	10 (0.48)	40 (1.9)
Second Floor	psf (kPa)	N/A	30 (1.4)	N/A	30 (1.4)
Roof (ground snow)	psf (kPa)	0	49 (2.3)	0	49 (2.3)
Attic	psf (kPa)	0	20 (0.96)	0	20 (0.96)
SEISMIC ZONE	----	2	2	2	2

Figure B1.2 Foundation Wall Loading Conditions and Building Geometry

1.3 ICF Above-Grade Wall Design

For above-grade wall construction, four different construction cases were investigated to cover the range of construction possibilities. The four construction cases are described below.

B1:

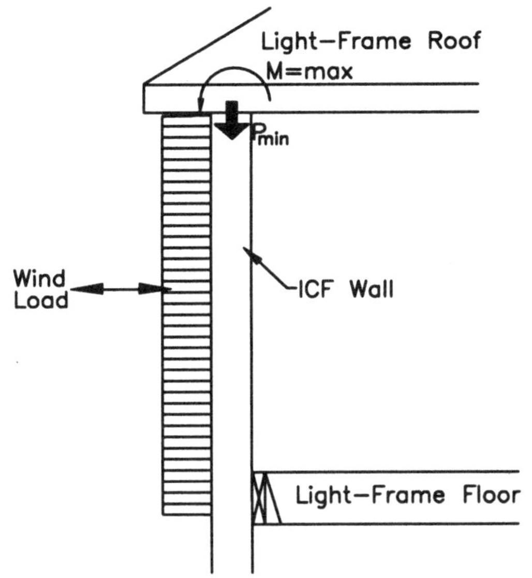

Supporting Light-Frame Roof Only

- Smallest axial design load with light-frame construction due to dead and live loads
- Largest practical eccentricity; assume roof truss bearing on wall with eccentricity
- Wind loads vary from 70 mph to 110 mph (113 to 177 km/hr)

B2:

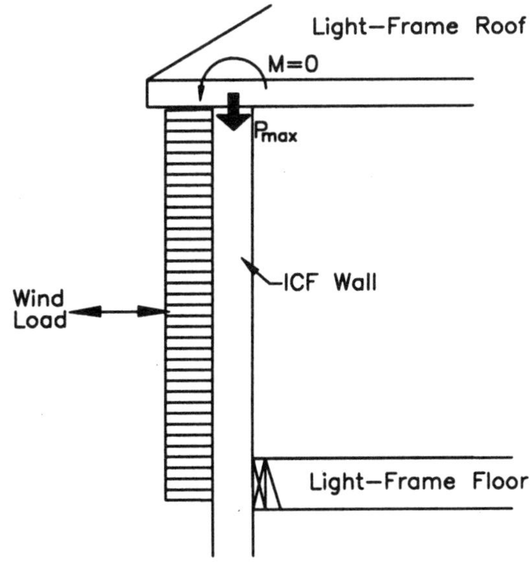

Supporting Light-Frame Roof Only

- Largest axial design load with light-frame construction due to dead and live loads
- No eccentricity; assume roof truss bearing on wall
- Wind loads vary from 70 mph to 110 mph (113 to 177 km/hr)

B3:

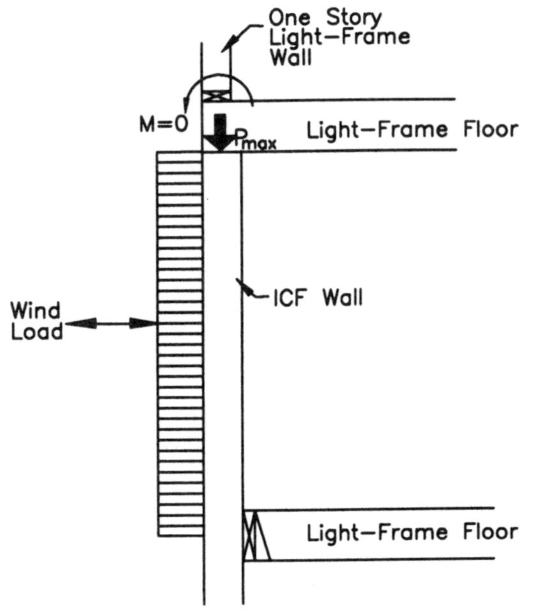

Supporting One-Story Light-Frame Construction and Roof

- Largest axial design load with light-frame construction due to dead and live loads
- No eccentricity; assume direct floor bearing on first-story wall
- Wind loads vary from 70 mph to 110 mph (113 to 177 km/hr)

B4:

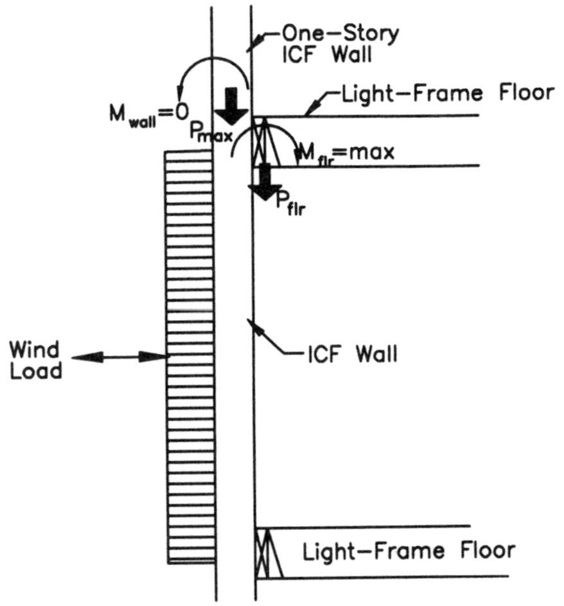

Supporting One-Story ICF Construction and Light-Frame Roof

- Largest axial design load with ICF construction due to dead and live loads
- Largest practical floor eccentricity; assume side-bearing ledger board for floor connection
- No wall eccentricity; assume ICF wall thickness for second-story wall is equal to ICF wall thickness for first-story wall
- Wind loads vary from 70 mph to 110 mph (113 to 177 km/hr)

The following values were used to design the above-grade walls; refer to Figure B1.3. The story wall height and wind load vary. The vertical wall reinforcements listed in the minimum vertical wall reinforcement tables in the *Prescriptive Method* are the results of calculations based on the governing construction case (B1 through B4) and ACI [B3] Load Combination.

Vinyl siding was considered as the exterior wall covering in all construction cases. In hindsight, the wall loading for construction cases B2, B3, and B4 should reflect a brick veneer; however, the lightest axial load case governed the minimum vertical wall reinforcement required. In addition, a brick veneer would induce a positive moment on the wall and counteract the negative moment resulting from the eccentric floor or roof loads as well as wind suction.

BUILDING GEOMETRY	Units	ABOVE-GRADE CONSTRUCTION CASE			
		B1	**B2**	**B3**	**B4**
Maximum Building Plan Dimension	feet (m)	60 (18)	60 (18)	60 (18)	60 (18)
Roof Slope	----	0:12	12:12	12:12	12:12
Roof Clear Span	feet (m)	40 (12)	40 (12)	40 (12)	40 (12)
First-Floor Clear Span	feet (m)	N/A	N/A	N/A	N/A
Second-Floor Clear Span	feet (m)	N/A	N/A	16 (4.9) to 32 (9.8)	16 (4.9) to 32 (9.8)
First-Story Wall Height	feet (m)	vary	vary	vary	vary
Second-Story Wall Height	feet (m)	N/A	N/A	10 (3)	10 (3)
First-Story Wall Thickness	inches (mm)	vary	vary	vary	vary
Second-Story Wall Thickness	inches (mm)	N/A	N/A	5.5 (140) light-frame	6 (152) ICF
Foundation Wall Thickness	psf (kPa)	N/A	N/A	N/A	N/A
DEAD LOADS					
First Floor	psf (kPa)	N/A	N/A	N/A	N/A
Second Floor	psf (kPa)	N/A	N/A	15 (0.72)	15 (0.72)
Roof and Ceiling	psf (kPa)	12 (0.58)	16 (0.77)	16 (0.77)	16 (0.77)
First-Story Wall	psf (kPa)	vary	vary	vary	vary
Second-Story Wall	psf (kPa)	N/A	N/A	11 (0.5)	75 (3.6)
Foundation Wall	psf (kPa)	N/A	N/A	N/A	N/A
LIVE LOADS					
First Floor	psf (kPa)	N/A	N/A	N/A	N/A
Second Floor	psf (kPa)	N/A	N/A	30 (1.4)	30 (1.4)
Roof (ground snow)	psf (kPa)	0	70 (3.3)	70 (3.3)	70 (3.3)
Attic	psf (kPa)	0	20 (0.96)	20 (0.96)	20 (0.96)
SEISMIC ZONE	----	2	2	2	2

Figure B1.3 Above-Grade Wall Loading Conditions and Building Geometry

1.4 ICF Lintel Design Assumptions

The moment and shear capacities were determined in accordance with ACI [B3]. The following design assumptions were used:

Second-Floor Live Load	=	30 *psf*	[1.4 *kPa*]
Attic Live Load	=	20 *psf*	[0.96 *kPa*]
Roof Dead Load	=	15 *psf*	[0.72 *kPa*]
Roof and Floor Clear Span	=	32 *ft*	[9.8 *m*]
ICF Wall Dead Load	=	69 *psf*	[3.3 *kPa*]
Deflection Criterion	=	*L*/240	

No. 3 *Stirrups are required at maximum d/*2, *where d = lintel depth minus* 2-*inch* (51-*mm) reinforcement cover.*

No. 4 *and No.* 5 *bars are used. All steel has a minimum tensile strength of* 40,000 *psi* (300 *MPa). All concrete has a minimum compressive strength,* f'_c, *of* 2,500 *psi* (17 *MPa).*

1.5 Ledger Board Connection Design Assumptions

The following design assumptions were used:

Floor Live Load	=	40 *psf*	[1.9 *kPa*]
Floor Dead Load	=	15 *psf*	[0.72 *kPa*]
Wind Load	=	40 *psf*	[1.9 *kPa*]
Wall Height above and below Connection	=	10 *feet*	[3 *m*]
Floor Joist Spacing	=	2 *feet*	[0.61 *m*]

2.0 PROPERTIES

2.1 Material Properties

2.1.1 Steel Reinforcement and Concrete

Assume the following minimums.

F_y = 40,000 *psi for steel reinforcement* [300 *MPa*]
f_c' = 2,500 *psi for concrete* [17 *MPa*]

2.1.2 Ledger Board

Assume a 1.5-inch x 7.25-inch (38-mm x 184-mm), No. 2 Grade Hem-Fir with the following allowable design properties determined in accordance with the *National Design Specification for Wood Construction (NDS)* [B4].

F_b = 850 *psi* [5.9 *MPa*]
F_b' = 850 *psi* (1.15) (1.6) (1.0)
= 1,564 *psi modified for flat use, wind load duration, and size* [10.8 *MPa*]
F_b' = 850 *psi* (1.0)(1.0)
= 850 *psi modified for live load duration and size* [5.9 *MPa*]
$F_{c\perp}$ = 405 *psi* [2.8 *MPa*]
$F_{c\perp}'$ = 405 *psi* (1.25) = 506 *psi modified for small bearing area* [3.5 *MPa*]
F_v = 75 *psi* [0.52 *MPa*]
F_v' = (75 *psi*) (2) = 150 *psi modified for no splits* [1.03 *MPa*]
S_{xx} = 13.14 in^3 [215 dm^3]
S_{yy} = 2.72 in^3 [44.6 dm^3]
h = 7.25 *inch* [184 *mm*]
b = 1.5 *inch* [38 *mm*]

2.1.3 Ledger Board Bolts

Assume A36 steel with the following design properties determined in accordance with the *Manual of Steel Construction Allowable Stress Design* [B5].

F_t = 19,100 *psi* [132 *MPa*]
F_y = 36,000 *psi* [248 *MPa*]
F_v = 9,860 *psi threads included in shear plane* [68 *MPa*]

2.2 Section Properties of Concrete

Due to the variety of available dimensions, a minimum equivalent rectangular section with the dimensions listed in Figure B2.1 is used to generate minimum reinforcement tables in the *Prescriptive Method* for use with as many ICF products as reasonably possible without adversely impacting design economy or practicality.

ICF Wall Type	Nominal Thickness inches (mm)	Minimum Equivalent Thickness inches (mm)	Minimum Equivalent Width inches (mm)	Vertical Core Spacing inches (mm)
Flat	3.5 (89)	3.5 (89)	12 (305)	N/A
	5.5 (140)	5.5 (140)	12 (305)	N/A
	7.5 (191)	7.5 (191)	12 (305)	N/A
	9.5 (241)	9.5 (241)	12 (305)	N/A
Waffle-Grid	6 (152)	5.0 (127)	6.25 (159)	12 (305)
	8 (203)	7.0 (178)	7.0 (178)	12 (305)
Screen-Grid	6 (152)	5.5 (140)	5.5 (140)	12 (305)

Figure B2.1 Equivalent Rectangular Section Dimensions

3.0 ICF FOUNDATION WALL DESIGN EXAMPLES AND ENGINEERING CALCULATIONS

The following engineering calculations are based on the application of several recognized engineering standards and specifications. The engineering methods were reviewed by a steering committee of industry experts and professional engineers under sponsorship of the U.S. Department of Housing and Urban Development (HUD), the Portland Cement Association (PCA), and the National Association of Home Builders (NAHB).

3.1 5.5-Inch- (140-mm-) Thick Flat ICF Basement Wall

For the purposes of illustration, a flat ICF foundation wall is selected from Table 3.4 of the *Prescriptive Method* for foundation walls constructed in soil with an equivalent fluid density of 30 pcf (481 kg/m^3). The foundation wall is 9 feet (2.7 m) high and has 5 feet (1.5 m) of unbalanced backfill. Table 3.4 shows that the foundation wall is required to have a minimum of one No. 4 bar at 48 inches (1.2 m) on center for vertical wall reinforcement. Calculate the capacity and check the adequacy of the 5.5-inch- (140-mm-) thick flat ICF foundation wall.

Using the values in Figure B1.2 and the material properties in Section 2.0, compute the amount of vertical wall reinforcement required in Construction Cases A1 through A4.

3.1.1 Construction Case A1

Figures B3.1 and B3.2 summarize the nominal and factored loads calculated in this section. Refer to Section 7.0 for typical loading condition equations. The wall moments are calculated by multiplying the axial load by the assumed eccentricities of the roof, wall above, and floor as applicable.

Dead Loads

First Floor	$0.5(32\ ft)(10\ psf) = 160\ plf$	[2.3 *N/m*]
Roof and Ceiling	$0.5(40\ ft)(12\ psf) = 240\ plf$	[3.5 *N/m*]
First-Story Wood Wall	$(8.0\ ft)(6\ psf) = 48\ plf$ @ *base of wall*	[0.70 *N/m*]
Foundation ICF Wall	$0.5(9.0\ ft)(68.75\ psf) = 309\ plf$ @ *mid-height*	[4.5 *N/m*]

Live Loads

First Floor	$0.5(32\ ft)(40 psf) = 640 plf$	[9.3 *N/m*]
Roof and Ceiling	$0.5(40\ ft)(0\ psf) = 0\ plf$	[0 *N/m*]
Attic	$0.5(40\ ft)(0\ psf) = 0\ plf$	[0 *N/m*]

Foundation Wall Moments

Dead Load$_{@top}$	$[240\ plf + 48\ plf + 160\ plf](0\ in) = 0\ in - lb/lf$	[0 *N-m/m*]
Live Load$_{@top}$	$(0\ plf + 0\ plf + 640\ plf)(0\ in) = 0\ in - lb/lf$	[0 *N-m/m*]

$$Earth\ Load_{@midht} = \frac{(30\ pcf)(5\ ft)(4.5\ ft)^2}{2} + \frac{(30\ pcf)(4.5\ ft)^3}{6} + 306\ plf(4.5\ ft) = 312\ ft-lb/lf \qquad [1.39\ kN\text{-}m/m]$$

$$Earth\ Load_{@x} = \frac{(30\ pcf)(5\ ft)(2.85\ ft)^2}{2} + \frac{(30\ pcf)(2.85\ ft)^3}{6} + 306\ plf(2.85\ ft) = 377\ ft-lb/lf \qquad [1.68\ kN\text{-}m/m]$$

x, 2.85 feet (0.87 m), is the location of the maximum moment; a simply supported beam model is used.

Parallel Shear

The unbalanced backfill height is greater than 4 feet (1.2 m). Assume all basement walls in the building have 5 feet (1.5 m) of unbalanced backfill height. Parallel shear is neglected since the foundation walls are generally restrained by the soil lateral pressure on all sides.

Perpendicular Shear

Refer to Section 7.0 for variable definitions.

$$V_{top} = \frac{ql^3}{6L} = \frac{(30\ pcf)(5ft)^3}{6(9ft)} = 69\ plf \qquad [1.0\ kN/m]$$

$$V_{midht} = qlx - \frac{qx^2}{2} - V_{bottom} = (30\ pcf)(5\ ft)(4.5\ ft) - \frac{(30\ pcf)(4.5\ ft)^2}{2} - 306\ plf = 66\ plf \qquad [0.96\ kN/m]$$

$$V_{bottom} = \frac{ql^2}{2} - V_{top} = \frac{(30\ pcf)(5\ ft)^2}{2} - 69\ plf = 306\ plf \qquad [4.46\ kN/m]$$

NOMINAL STRUCTURAL LOAD SUMMARY

Story	Vertical Location within Wall	Dead		Live		Wind			Earth		
		Axial (plf)	Moment (in-lb/lf)	Axial (plf)	Moment (in-lb/lf)	Moment (in-lb/lf)	Shear,$_{perp}$ (plf)	Shear,$_{par}$ (lb)	Moment (in-lb/lf)	Shear,$_{perp}$ (plf)	Shear,$_{par}$ (lb)
Foundation	Top	448	0	640	0				0	69	
	Midheight	757	0	640	0				3,745	66	
	x = 2.85 ft	871	0	640	0				4,530	0	
	Bottom	1,067	0	640	0				0	306	0

For SI: 1 plf = 14.59390 N/m; 1 in-lb/lf = 0.3706850 N-m/m; 1 lb = 4.448222 N

Figure B3.1 Nominal Load Summary for Construction Case A1

Figure B3.1 is a summary sheet of the loading conditions described above. To determine the moments at various locations in each wall story, refer to Section 7.0. Figure B3.2 is a summary sheet of the ACI [B3] factored loads for each ICF wall story. The values listed in Figure B3.2 are determined by substituting the values from Figure B3.1 into the equations listed in the left column of Figure B3.2.

FACTORED STRUCTURAL LOAD SUMMARY												
FOUNDATION WALL												
		Factored Loads										
ACI 318 Load Combinations[1]	Vertical Location within Wall	Dead		Live		Earth			Total			
		Axial (plf)	Moment (in-lb/lf)	Axial (plf)	Moment (in-lb/lf)	Moment (in-lb/lf)	Shear,perp (plf)	Shear,par (lb)	Axial (plf)	Moment (in-lb/lf)	Shear,perp (plf)	Shear,par (lb)
1. U = 1.4D + 1.7L	Top	627	0	1,088	0				1,715	0		
	Midheight	1,060	0	1,088	0				2,148	0		
	x = 2.85 ft	1,219	0	1,088	0				2,307	0		
	Bottom	1,494	0	1,088	0				2,582	0		
2. U = 1.4D + 1.7L + 1.7H	Top	627	0	1,088	0	0	118	N/A	1,715	0	118	N/A
	Midheight	1,060	0	1,088	0	6,367	112		2,148	6,367	112	
	x = 2.85 ft	1,219	0	1,088	0	7,701	0		2,307	7,701	0	
	Bottom	1,494	0	1,088	0	0	520		2,582	0	520	
3. U = 0.9D + 1.7H	Top	403	0			0	118	N/A	403	0	118	N/A
	Midheight	682	0			6,367	112		682	6,367	112	
	x =2.85 ft	784	0			7,701	0		784	7,701	0	
	Bottom	960	0			0	520		960	0	520	

[1] D = Dead Load, L = Live Load, W = Wind Load, H = Earth Load

For SI: 1 plf = 14.59390 N/m; 1 in-lb/lf = 0.3706850 N-m/m; 1 lb = 4.448222 N

Figure B3.2 Factored Load Summary for Construction Case A1

Assume the section does not require reinforcement.

Check Perpendicular Shear

According to Figure B3.2, the critical factored perpendicular shear load, V_u, experienced by the foundation wall occurs at the bottom of the wall story due to ACI [B3] Load Combinations (2) and (3).

$$V_u = 520 \ plf \qquad [7.6 \ kN/m]$$

$$\phi V_n = 0.65\left(\frac{4}{3}\right)\sqrt{f'_c}\, b_w h = 0.65\left(\frac{4}{3}\right)\sqrt{2{,}500 \ psi}(12 \ in)(5.5 \ in) = 2{,}860 \ plf \qquad [41.7 \ kN/m]$$

$$V_u \le \phi V_n \quad OK$$

Check Compression and Tension

According to Figure B3.2, the critical maximum total moment, M_u, experienced by the foundation story wall occurs at x = 2.85 feet (0.9 m) due to ACI [B3] Load Combinations (2) and (3). The corresponding total factored axial load, P_u, is also taken at x = 2.85 feet (0.9 m) of the foundation wall based on ACI [B3] Load Combinations (2) and (3). The minimum factored moment required for design is specified in ACI 22.6.3 [B3].

$$M_u = 7{,}701 \ in-lb/ft = 0.64 \ ft-kip/lf \longleftarrow GOVERNS \qquad [2.9 \ kN\text{-}m/m]$$

$$P_u = 2{,}307 \ plf = 2.3 \ kip \ per \ ft \ (ACI \ LOAD \ COMBINATION \ 2) \qquad [34 \ kN/m]$$

$$P_u = 784 \ plf = 0.78 \ kip \ per \ ft \ (ACI \ LOAD \ COMBINATION \ 3) \qquad [11 \ kN/m]$$

$$M_{u,min} = 0.1hP_u = 0.1(5.5 \ in)(2{,}307 \ plf) = 1{,}269 \ in-lb/lf = 0.11 \ ft-kip/lf \ (ACI \ LOAD \ COMBINATION \ 2) \qquad [470 \ N\text{-}m/m]$$

$$M_{u,min} = 0.1hP_u = 0.1(5.5\,in)(784\,plf) = 431\,in-lb/lf$$
$$= 0.04\,ft-kip/lf\;(ACI\;\;LOAD\;\;COMBINATION\;3)$$ [160 *N-m/m*]

Plot M_u and P_u on the interaction diagram for a 5.5-inch- (140-mm-) thick flat wall shown in Figure B3.3. The plotted point, "A1", lies within the lower tension boundary, the upper compression boundary for a 9-foot (2.7-m) wall height, and the reference axes; therefore, a 5.5-inch- (140-mm-) thick flat structural plain concrete wall is sufficient for the given loading conditions.

Check Deflection

For below-grade walls, a conservative deflection limit of *L/240* for service live loads is used because earth loads are immediate and are not expected to change significantly with time. To calculate wall deflection at service load levels, effective section properties of the assumed uncracked concrete section are based on E_cI_g.

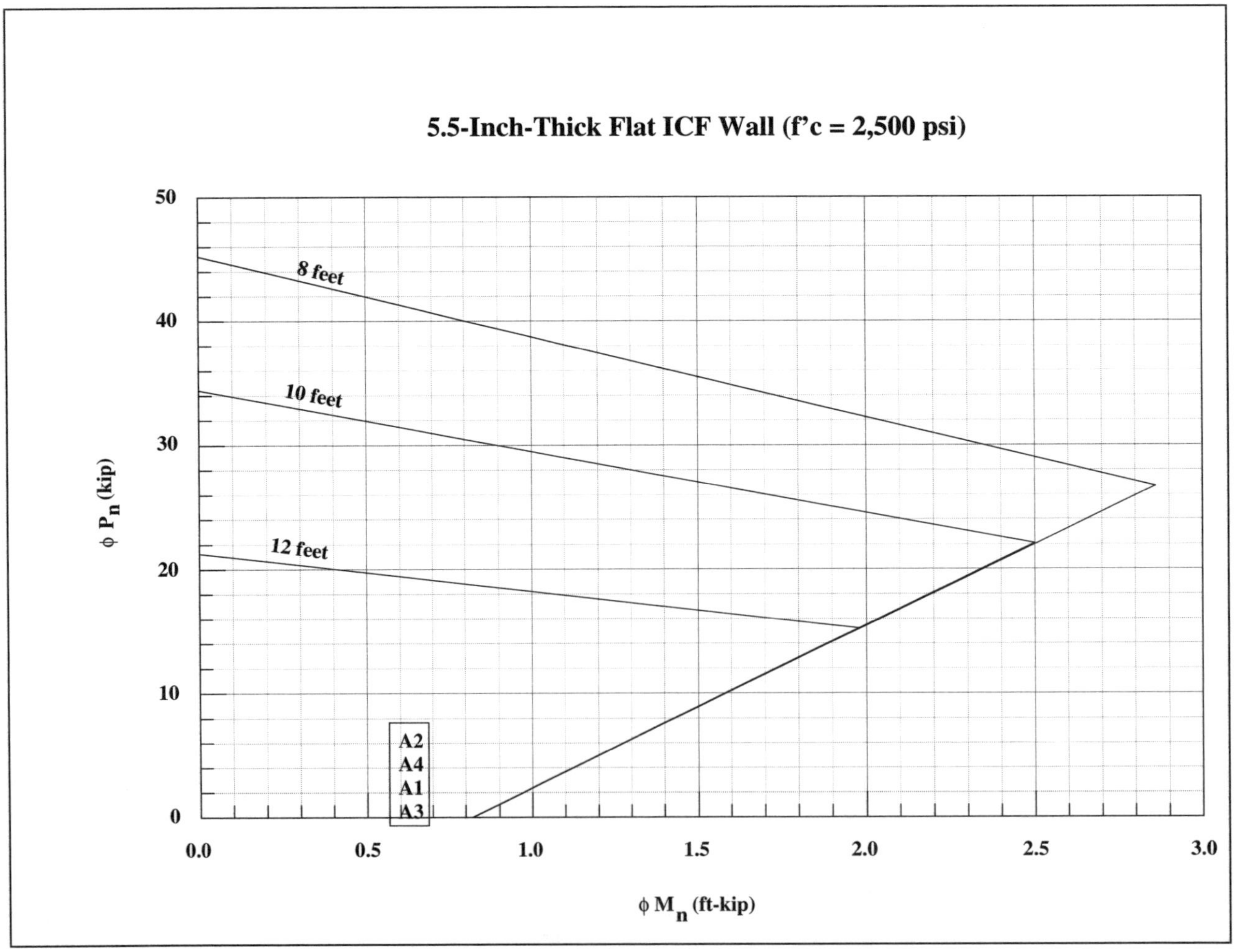

Figure B3.3 Structural Plain Concrete Wall Interaction Diagram

Assume a gypsum board interior finish exposed to view. For simplicity, when calculating the maximum deflection, assume the earth load acts on the entire wall height.

$$\Delta_{maximum} = \frac{0.01304(0.5)ql^5}{E_c I_g} = \frac{0.01304(0.5)(30\,pcf)(1\,ft)(9\,ft)^5}{(2{,}850{,}000\,psi)\left(\frac{(5.5\,in)(12\,in)^3}{12}\right)}\left(\frac{1{,}728\,in^3}{ft^3}\right) = 0.009\,in \qquad [0.2\,mm]$$

$$\Delta_{allowable} = \frac{L}{240} = \frac{(9\,ft)\left(\frac{12\,in}{ft}\right)}{240} = 0.45\,in \qquad [11\,mm]$$

$$\Delta_{actual} \le \Delta_{allowable} \qquad OK$$

Determine Reinforcement

Although the wall is designed as a structural plain concrete wall, a nominal amount of reinforcement is typically specified. Tests [B6] have shown that horizontal and vertical wall reinforcement spacing limited to 48 inches (1.2 m) on center results in reliable performance; therefore, install one Grade 40 (300 MPa), No. 4 bar at 48 inches (1.2 m) on center for minimum vertical wall reinforcement and one No. 4 bar at 48 inches (1.2 m) on center for minimum horizontal wall reinforcement. At least one continuous horizontal reinforcement bar should be placed within the top 12 inches (305 mm) of the wall story.

3.1.2 Construction Case A2

Figures B3.4 and B3.5 summarize the nominal and factored loads calculated in this section. Refer to Section 7.0 for typical loading condition equations. The wall moments are calculated by multiplying the axial load by the assumed eccentricities of the roof, wall above, and floor as applicable.

Dead Loads		
First Floor	$0.5(32\,ft)(15\,psf) = 240\,plf$	[3.5 *kN/m*]
Second Floor	$0.5(32\,ft)(15\,psf) = 240\,plf$	[3.5 *kN/m*]
Roof and Ceiling	$0.5(40\,ft)(16\,psf) = 320\,plf$	[4.7 *kN/m*]
First-Story Wood Wall	$(10\,ft)(11\,psf) = 110\,plf$ *@ base of wall*	[1.6 *kN/m*]
Second-Story Wood Wall	$(10\,ft)(11\,psf) = 110\,plf$ *@ base of wall*	[1.6 *kN/m*]
Foundation ICF Wall	$0.5(9.0\,ft)(68.75\,psf) = 309\,plf$ *@ mid-height*	[4.5 *kN/m*]
Live Loads		
First Floor	$0.5(32\,ft)(40\,psf) = 640\,plf$	[9.3 *kN/m*]
Second Floor	$0.5(32\,ft)(30\,psf) = 480\,plf$	[7 *kN/m*]
Roof and Ceiling	$0.5(40\,ft)(49\,psf) = 980\,plf$	[14.3 *kN/m*]
Attic	$0.5(40\,ft)(20\,psf) = 400\,plf$	[5.8 *kN/m*]

Foundation Wall Moments

Dead Load $_{@top}$ $\left[2(240)\,plf + 2(110)\,plf + 320\,plf\right](0\,in) = 0\,in - lb/lf$ [0 *N-m/m*]

Live Load $_{@top}$ $\left[(640\,plf + 480\,plf + 400\,plf + 980\,plf)\right](0\,in) = 0\,in - lb/lf$ [0 *N-m/m*]

Earth Load *Unchanged from Construction Case* A1; *refer to Section* 3.1.1.

Parallel Shear

Unchanged from Construction Case A1; *refer to Section* 3.1.1.

Perpendicular Shear

Unchanged from Construction Case A1; *refer to Section* 3.1.1.

Figure B3.4 is a summary sheet of the loading conditions described above. To determine the moments at various locations in each wall story, refer to Section 7.0. Figure B3.5 is a summary sheet of the ACI [B3] factored loads for each ICF wall story. The values listed in Figure B3.5 are determined by substituting the values from Figure B3.4 into the equations listed in the left column of Figure B3.5.

NOMINAL STRUCTURAL LOAD SUMMARY

Story	Vertical Location within Wall	Dead		Live		Wind			Earth		
		Axial (plf)	Moment (in-lb/lf)	Axial (plf)	Moment (in-lb/lf)	Moment (in-lb/lf)	Shear$_{,perp}$ (plf)	Shear$_{,par}$ (lb)	Moment (in-lb/lf)	Shear$_{,perp}$ (plf)	Shear$_{,par}$ (lb)
Foundation	Top	1,020	0	2,500	0				0	69	
	Midheight	1,329	0	2,500	0				3,745	66	
	x = 2.85 ft	1,443	0	2,500	0				4,530	0	
	Bottom	1,639	0	2,500	0				0	306	N/A

For SI: 1 plf = 14.59390 N/m; 1 in-lb/lf = 0.3706850 N-m/m; 1 lb = 4.448222 N

Figure B3.4 Nominal Load Summary for Construction Case A2

FACTORED STRUCTURAL LOAD SUMMARY

FOUNDATION WALL

ACI 318 Load Combinations[1]	Vertical Location within Wall	Factored Loads										
		Dead		Live		Earth			Total			
		Axial (plf)	Moment (in-lb/lf)	Axial (plf)	Moment (in-lb/lf)	Moment (in-lb/lf)	Shear$_{,perp}$ (plf)	Shear$_{,par}$ (lb)	Axial (plf)	Moment (in-lb/lf)	Shear$_{,perp}$ (plf)	Shear$_{,par}$ (lb)
1. U = 1.4D + 1.7L	Top	1,428	0	4,250	0				5,678	0		
	Midheight	1,861	0	4,250	0				6,111	0		
	x = 2.85 ft	2,020	0	4,250	0				6,270	0		
	Bottom	2,295	0	4,250	0				6,545	0		
2. U = 1.4D + 1.7L + 1.7H	Top	1,428	0	4,250	0	0	118	N/A	5,678	0	118	N/A
	Midheight	1,861	0	4,250	0	6,367	112		6,111	6,367	112	
	x = 2.85 ft	2,020	0	4,250	0	7,701	0		6,270	7,701	0	
	Bottom	2,295	0	4,250	0	0	520		6,545	0	520	
3. U = 0.9D + 1.7H	Top	918	0			0	118	N/A	918	0	118	N/A
	Midheight	1,196	0			6,367	112		1,196	6,367	112	
	x = 2.85 ft	1,299	0			7,701	0		1,299	7,701	0	
	Bottom	1,475	0			0	520		1,475	0	520	

[1] D = Dead Load, L = Live Load, W = Wind Load, H = Earth Load

For SI: 1 plf = 14.59390 N/m; 1 in-lb/lf = 0.3706850 N-m/m; 1 lb = 4.448222 N

Figure B3.5 Factored Load Summary for Construction Case A2

Assume the section does not require reinforcement.

Check Perpendicular Shear

According to Figure B3.5, the critical factored perpendicular shear load, V_u, is the same as that in Construction Case A1; refer to Section 3.1.1 for calculations.

Check Compression and Tension

According to Figure B3.5, the critical maximum total moment, M_u, experienced by the foundation story wall occurs at x = 2.85 feet (0.87 m) due to ACI [B3] Load Combinations (2) and (3). The corresponding total factored axial load, P_u, is also taken at x = 2.85 feet (0.87 m) of the foundation wall based on ACI [B3] Load Combinations (2) and (3). The minimum factored moment required for design is specified in ACI 22.6.3 [B3].

$$M_u = 7{,}701\,in-lb/lf = 0.64\,ft-kip/lf \longleftarrow GOVERNS \qquad [2.85\ kN\text{-}m/m]$$
$$P_u = 6{,}270\,plf = 6.3\,kip\ per\ ft\ (ACI\ LOAD\ COMBINATION\ 2) \qquad [92\ kN/m]$$
$$P_u = 1{,}299\,plf = 1.3\,kip\ per\ ft\ (ACI\ LOAD\ COMBINATION\ 3) \qquad [19\ kN/m]$$
$$M_{u,min} = 0.1hP_u = 0.1(5.5\,in)(6{,}270\,plf) = 3{,}449\,in-lb/lf = 0.29\,ft-kip/lf\ (ACI\ LOAD\ COMBINATION\ 2) \qquad [1.3\ kN\text{-}m/m]$$
$$M_{u,min} = 0.1hP_u = 0.1(5.5\,in)(1{,}299\,plf) = 715\,in-lb/lf = 0.06\,ft-kip/lf\ (ACI\ LOAD\ COMBINATION\ 3) \qquad [0.26\ kN\text{-}m/m]$$

Plot M_u and P_u on the interaction diagram for a 5.5-inch- (140-mm-) thick flat wall shown in Figure B3.3. The plotted point, "A2", lies within the lower tension boundary, the upper compression boundary for a 9-foot (2.7-m) wall height, and the reference axes; therefore, a 5.5-inch- (140-mm-) thick flat wall structural plain concrete wall is sufficient for the given loading conditions.

Check Deflection

Unchanged from Construction Case A1; refer to Section 3.1.1 for calculations.

Determine Reinforcement

Unchanged from Construction Case A1; refer to Section 3.1.1 for calculations.

3.1.3 Construction Case A3

Figures B3.6 and B3.7 summarize the nominal and factored loads calculated in this section. Refer to Section 7.0 for typical loading condition equations. The wall moments are calculated by multiplying the axial load by the assumed eccentricities of the roof, wall above, and floor as applicable.

Dead Loads

First Floor	$0.5(32\,ft)(10\,psf) = 160\,plf$	[2.3 *kN/m*]
Roof and Ceiling	$0.5(40\,ft)(12\,psf) = 240\,plf$	[3.5 *kN/m*]
First-Story ICF Wall	$0.5(8\,ft)(50\,psf) = 200\,plf$ @ *mid-height*	[2.9 *kN/m*]
Foundation ICF Wall	$0.5(9.0\,ft)(68.75\,psf) = 309\,plf$ @ *mid-height*	[4.5 *kN/m*]

Live Loads

First Floor	$0.5(32\,ft)(10\,psf) = 160\,plf$	[2.3 *kN/m*]
Roof and Ceiling	$0.5(40\,ft)(0\,psf) = 0\,plf$	[0 *kN/m*]
Attic	$0.5(40\,ft)(0\,psf) = 0\,plf$	[0 *kN/m*]

Foundation Wall Moments

Dead Load $_{@top}$	$[240\,plf + 2(200\,plf)](0.75in) + (160\,plf)(-2.75in) = 40\,in-lb/lf$	[15 *N-m/m*]
Live Load $_{@top}$	$(0\,plf + 0\,plf)(0.75in) + (160\,plf)(-2.75in) = -440\,in-lb/lf$	[-163 *N-m/m*]
Earth Load	*Unchanged from Construction Case A1; refer to Section* 3.1.1.	

Parallel Shear

Unchanged from Construction Case A1; refer to Section 3.1.1.

Perpendicular Shear

Unchanged from Construction Case A1; refer to Section 3.1.1.

NOMINAL STRUCTURAL LOAD SUMMARY

Story	Vertical Location within Wall	Dead		Live		Wind			Earth		
		Axial (plf)	Moment (in-lb/lf)	Axial (plf)	Moment (in-lb/lf)	Moment (in-lb/lf)	$Shear_{,perp}$ (plf)	$Shear_{,par}$ (lb)	Moment (in-lb/lf)	$Shear_{,perp}$ (plf)	$Shear_{,par}$ (lb)
Foundation	Top	800	40	160	-440				0	69	
	Midheight	1,109	20	160	-220				3,745	66	
	x = 2.85 ft	1,223	13	160	-139				4,530	0	
	Bottom	1,419	0	160	0				0	306	N/A

For SI: 1 plf = 14.59390 N/m; 1 in-lb/lf = 0.3706850 N-m/m; 1 lb = 4.448222 N

Figure B3.6 Nominal Load Summary for Construction Case A3

Figure B3.6 is a summary sheet of the loading conditions described above. To determine the moments at various locations in each wall story, refer to Section 7.0. Figure B3.7 is a summary sheet of the ACI [B3] factored loads for each ICF wall story. The values listed in Figure B3.7 are determined by substituting the values from Figure B3.6 into the equations listed in the left column of Figure B3.7.

Check Perpendicular Shear

According to Figure B3.7, the critical factored perpendicular shear load, V_u, is the same as that in Construction Case A1; refer to Section 3.1.1 for calculations.

FACTORED STRUCTURAL LOAD SUMMARY												
FOUNDATION WALL												
		Factored Loads										
ACI 318 Load Combinations[1]	Vertical Location within Wall	Dead		Live		Earth			Total			
		Axial (plf)	Moment (in-lb/lf)	Axial (plf)	Moment (in-lb/lf)	Moment (in-lb/lf)	$Shear_{,perp}$ (plf)	$Shear_{,par}$ (lb)	Axial (plf)	Moment (in-lb/lf)	$Shear_{,perp}$ (plf)	$Shear_{,par}$ (lb)
1. U = 1.4D + 1.7L	Top	1,120	56	272	-748				1,392	-692		
	Midheight	1,553	28	272	-374				1,825	-346		
	x = 2.85 ft	1,712	18	272	-237				1,984	-219		
	Bottom	1,986	0	272	0				2,258	0		
2. U = 1.4D + 1.7L + 1.7H	Top	1,120	56	272	-748	0	118	N/A	1,392	-692	118	N/A
	Midheight	1,553	28	272	-374	6,367	112		1,825	6,021	112	
	x = 2.85 ft	1,712	18	272	-237	7,701	0		1,984	7,482	0	
	Bottom	1,986	0	272	0	0	520		2,258	0	520	
3. U = 0.9D + 1.7H	Top	720	36			0	118	N/A	720	36	118	N/A
	Midheight	998	18			6,367	112		998	6,385	112	
	x = 2.85 ft	1,101	11			7,701	0		1,101	7,712	0	
	Bottom	1,277	0			0	520		1,277	0	520	

[1] D = Dead Load, L = Live Load, W = Wind Load, H = Earth Load

For SI: 1 plf = 14.59390 N/m; 1 in-lb/lf = 0.3706850 N-m/m; 1 lb = 4.448222 N

Figure B3.7 Factored Load Summary for Construction Case A3

Assume the section does not require reinforcement.

Check Compression and Tension

According to Figure B3.7, the critical maximum total moment, M_u, experienced by the foundation story wall occurs at x = 2.85 feet (0.87 m) due to ACI [B3] Load Combination (3). The corresponding total factored axial load, P_u, is also taken at x = 2.85 feet (0.87 m) of the foundation wall based on ACI [B3] Load Combination (3). The minimum factored moment required for design is specified in ACI 22.6.3 [B3].

$$M_u = 7{,}712\, in-lb/lf = 0.64\, ft-kip/lf \longleftarrow GOVERNS \quad [2.9\ kN\text{-}m/m]$$

$$P_u = 1{,}101\, plf = 1.1\ kip/lf \quad [16\ kN/m]$$

$$M_{u,min} = 0.1hP_u = 0.1(5.5\, in)(1{,}101\, plf) = 605.6\, in-lb/lf = 0.05\ ft-kip/lf \quad [0.2\ kN\text{-}m/m]$$

Plot M_u and P_u on the interaction diagram for a 5.5-inch- (140-mm-) thick flat wall shown in Figure B3.3. The plotted point, “A3”, lies within the lower tension boundary, the upper compression boundary for a 9-foot (2.7-m) wall height, and the reference axes; therefore, a 5.5-inch- (140-mm-) thick flat wall structural plain concrete wall is sufficient for the given loading conditions.

Check Deflection

Unchanged from Construction Case A1; refer to Section 3.1.1 for calculations.

Determine Reinforcement

Unchanged from Construction Case A1; refer to Section 3.1.1 for calculations.

3.1.4 Construction Case A4

Figures B3.8 and B3.9 summarize the nominal and factored loads calculated in this section. Refer to Section 7.0 for typical loading condition equations. The wall moments are calculated by multiplying the axial load by the assumed eccentricities of the roof, wall above, and floor as applicable.

Dead Loads

First Floor	$0.5(32\ ft)(15\ psf) = 240\ plf$	[3.5 *kN/m*]
Second Floor	$0.5(32\ ft)(15\ psf) = 240\ plf$	[3.5 *kN/m*]
Roof and Ceiling	$0.5(40\ ft)(16\ psf) = 320\ plf$	[4.7 *kN/m*]
First-Story ICF Wall	$0.5(10\ ft)(100\ psf) = 500\ plf$ *@ mid-height*	[7.3 *kN/m*]
Second-Story ICF Wall	$0.5(10\ ft)(75\ psf) = 375\ plf$ *@ mid-height*	[5.5 *kN/m*]
Foundation ICF Wall	$0.5(9\ ft)(68.75\ psf) = 309\ plf$ *@ mid-height*	[4.5 *kN/m*]

Live Loads

First Floor	$0.5(32\ ft)(40\ psf) = 640\ plf$	[9.3 *kN/m*]
Second Floor	$0.5(32\ ft)(30\ psf) = 480\ plf$	[7 *kN/m*]
Roof and Ceiling	$0.5(40\ ft)(49\ psf) = 980\ plf$	[14.3 *kN/m*]
Attic	$0.5(40\ ft)(20\ psf) = 400\ plf$	[5.8 *kN/m*]

Foundation Wall Moments

Dead Load$_{@top}$ $[240\ plf + 2(375\ plf) + 2(500\ plf) + 320\ plf](0\ in)$
$+[240\ plf(-4.75\ in)] = -1{,}140\ in-lb/lf$ [-0.42 *N-m/m*]

Live Load$_{@top}$ $[(480\ plf + 400\ plf + 980\ plf)](0\ in) + [640\ plf(-4.75\ in)]$
$= -3{,}040\ in-lb/lf$ [-1.1 *kN-m/m*]

Earth Load *Unchanged from Construction Case A1; refer to Section* 3.1.1.

Parallel Shear

Unchanged from Construction Case A1; refer to Section 3.1.1.

Perpendicular Shear

Unchanged from Construction Case A1; refer to Section 3.1.1.

NOMINAL STRUCTURAL LOAD SUMMARY											
Story	Vertical Location within Wall	Dead		Live		Wind			Earth		
		Axial (plf)	Moment (in-lb/lf)	Axial (plf)	Moment (in-lb/lf)	Moment (in-lb/lf)	$Shear_{,perp}$ (plf)	$Shear_{,par}$ (lb)	Moment (in-lb/lf)	$Shear_{,perp}$ (plf)	$Shear_{,par}$ (lb)
Foundation	Top	2,550	-1,140	2,500	-3,040				0	69	
	Midheight	2,859	-570	2,500	-1,520				3,745	66	
	x = 2.85 ft	2,973	-361	2,500	-962				4,530	0	
	Bottom	3,169	0	2,500	0				0	306	N/A

For SI: 1 plf = 14.59390 N/m; 1 in-lb/lf = 0.3706850 N-m/m; 1 lb = 4.448222 N

Figure B3.8 Nominal Load Summary for Construction Case A4

FACTORED STRUCTURAL LOAD SUMMARY												
FOUNDATION WALL												
ACI 318 Load Combinations[1]	Vertical Location within Wall	Factored Loads										
		Dead		Live		Earth			Total			
		Axial (plf)	Moment (in-lb/lf)	Axial (plf)	Moment (in-lb/lf)	Moment (in-lb/lf)	$Shear_{,perp}$ (plf)	$Shear_{,par}$ (lb)	Axial (plf)	Moment (in-lb/lf)	$Shear_{,perp}$ (plf)	$Shear_{,par}$ (lb)
1. U = 1.4D + 1.7L	Top	3,570	-1,596	4,250	-5,168				7,820	-6,764		
	Midheight	4,003	-798	4,250	-2,584				8,253	-3,382		
	x = 2.85 ft	4,162	-505	4,250	-1,636				8,412	-2,141		
	Bottom	4,436	0	4,250	0				8,686	0		
2. U = 1.4D + 1.7L + 1.7H	Top	3,570	-1,596	4,250	-5,168	0	118	N/A	7,820	-6,764	118	N/A
	Midheight	4,003	-798	4,250	-2,584	6,367	112		8,253	2,985	112	
	x = 2.85 ft	4,162	-505	4,250	-1,636	7,701	0		8,412	5,560	0	
	Bottom	4,436	0	4,250	0	0	520		8,686	0	520	
3. U = 0.9D + 1.7H	Top	2,295	-1,026			0	118	N/A	2,295	-1,026	118	N/A
	Midheight	2,573	-513			6,367	112		2,573	5,854	112	
	x = 2.85 ft	2,676	-325			7,701	0		2,676	7,376	0	
	Bottom	2,852	0			0	520		2,852	0	520	

[1] D = Dead Load, L = Live Load, W = Wind Load, H = Earth Load

For SI: 1 plf = 14.59390 N/m; 1 in-lb/lf = 0.3706850 N-m/m; 1 lb = 4.448222 N

Figure B3.9 Factored Load Summary for Construction Case A4

Figure B3.8 is a summary sheet of the loading conditions described above. To determine the moments at various locations in each wall story, refer to Section 7.0. Figure B3.9 is a summary sheet of the ACI [B3] factored loads for each ICF wall story. The values listed in Figure B3.9 are determined by substituting the values from Figure B3.8 into the equations listed in the left column of Figure B3.9.

Assume the section does not require reinforcement.

Check Perpendicular Shear

According to Figure B3.9, the critical factored perpendicular shear load, V_u, is the same as that in Construction Case A1; refer to Section 3.1.1 for calculations.

Check Compression and Tension

According to Figure B3.9, the critical maximum total moment, M_u, experienced by the foundation story wall occurs at x = 2.85 feet (0.87 m) due to ACI [B3] Load Combination (3). The corresponding total factored axial load, P_u, is also taken at x = 2.85 feet (0.87 m) of the foundation wall based on ACI [B3] Load Combination (3). The minimum factored moment required for design is specified in ACI 22.6.3 [B3].

$$M_u = 7{,}376\, in - lb/ft = 0.61\, ft - kip/lf \longleftarrow GOVERNS \qquad [2.7\ kN\text{-}m/m]$$

$$P_u = 2{,}676\, plf = 2.7\, kip/lf \qquad [39\ kN/m]$$

$$M_{u,min} = 0.1hP_u = 0.1(5.5\, in)(2{,}676\, plf) = 1{,}472\, in - lb/lf = 0.12\, ft - kip/lf \qquad [0.55\ kN\text{-}m/m]$$

Plot M_u and P_u on the interaction diagram for a 5.5-inch- (140-mm-) thick flat wall shown in Figure B3.3. The plotted point, "A4", lies within the lower tension boundary, the upper compression boundary for a 9-foot (2.7-m) wall height, and the reference axes; therefore, a 5.5-

inch- (140-mm-) thick flat wall structural plain concrete wall is sufficient for the given loading conditions.

Check Deflection

Unchanged from Construction Case A1; refer to Section 3.1.1 for calculations.

Determine Reinforcement

Unchanged from Construction Case A1; refer to Section 3.1.1 for calculations.

3.1.5 Construction Case Summary

In Construction Cases A1 through A4, a 5.5-inch- (140-mm-) thick flat ICF basement wall, 9 feet (2.7 m) high with 5 feet (1.5 m) of unbalanced backfill with an equivalent fluid density of 30 pcf (481 kg/m^3), requires one No. 4 bar at 48 inches (1.2 m) on center minimum vertical wall reinforcement and one No. 4 bar at 48 inches (1.2 m) on center minimum horizontal wall reinforcement. The wall reinforcement is not required by calculation, but it is a recommendation based on reliable practice to control temperature and shrinkage cracking and potential construction error. From Table 3.3 of the *Prescriptive Method*, we also obtain one No. 4 horizontal bar near third points in the wall story (approximately 3 feet (0.9 m) on center) and, from Table 3.4 of the *Prescriptive Method*, one No. 4 bar at 48 inches (1.2 m) on center minimum vertical wall reinforcement.

TABLE 3.3
MINIMUM HORIZONTAL WALL REINFORCEMENT FOR ICF BASEMENT WALLS
(excerpt from the *Prescriptive Method*)

Maximum Height of Basement Wall feet (meters)	Location of Horizontal Reinforcement
8 (2.4)	One No. 4 bar within 12 inches (305 mm) of the top of the wall story and one No. 4 bar near mid-height of the wall story
9 (2.7)	**One No. 4 bar within 12 inches (305 mm) of the top of the wall story and one No. 4 bar near third points in the wall story**

TABLE 3.4
MINIMUM VERTICAL WALL REINFORCEMENT FOR 5.5-INCH- (140-MM-) THICK FLAT ICF BASEMENT WALLS [1,2,3]

(excerpt from the *Prescriptive Method*)

Max. Height of Basement Wall (feet)	Maximum Unbalanced Backfill Height[4] (feet)	Minimum Vertical Reinforcement		
		Maximum Equivalent Fluid Density 30 pcf	Maximum Equivalent Fluid Density 45 pcf	Maximum Equivalent Fluid Density 60 pcf
8	4	#4@48"	#4@48"	#4@48"
	5	#4@48"	#3@12"; #4@22"; #5@32"; #6@40"	#3@8"; #4@14"; #5@20"; #6@26"
	6	#3@12"; #4@22"; #5@30"; #6@40"	#3@8"; #4@14"; #5@20"; #6@24"	#3@6"; #4@10"; #5@14"; #6@20"
	7	#3@8"; #4@14"; #5@22"; #6@26"	#3@5"; #4@10"; #5@14"; #6@18"	#3@4"; #4@6"; #5@10"; #6@14"
9	4	#4@48"	#4@48"	#4@48"
	5	**#4@48"**	#3@12"; #4@20"; #5@28"; #6@36"	#3@8"; #4@14"; #5@20"; #6@22"
	6	#3@10"; #4@20"; #5@28"; #6@34"	#3@6"; #4@12"; #5@18"; #6@20"	#4@8"; #5@14"; #6@16"
	7	#3@8"; #4@14"; #5@20"; #6@22"	#4@8"; #5@12"; #6@16"	#4@6"; #5@10"; #6@12"
	8	#3@6"; #4@10"; #5@14"; #6@16"	#4@6"; #5@10"; #6@12"	#4@4"; #5@6"; #6@8"
10	4	#4@48"	#4@48"	#4@48"
	5	#4@48"	#3@10"; #4@18"; #5@26"; #6@30"	#3@6"; #4@14"; #5@18"; #6@20"
	6	#3@10"; #4@18"; #5@24"; #6@30"	#3@6"; #4@12"; #5@16"; #6@18"	#3@4"; #4@8"; #5@12"; #6@14"
	7	#3@6"; #4@12"; #5@16"; #6@18"	#3@4"; #4@8"; #5@12	#4@6"; #5@8"; #6@10"
	8	#3@4"; #4@8"; #5@12"; #6@14"	#4@6"; #5@8"; #6@12"	#4@4"; #5@6"; #6@8"
	9	#3@4"; #4@6"; #5@10"; #6@12"	#4@4"; #5@6"; #6@8"	#5@4"; #6@6"

For SI: 1 foot = 0.3048 m; 1 inch = 25.4 mm; 1 pcf = 16.0179 kg/m^3

1 Deflection criterion is *L/240,* where *L* is the height of the basement wall in inches.
2 Interpolation shall not be permitted.
3 Walls shall be laterally supported at the top before backfilling.
4 Refer to Section 1.0 for the definition of unbalanced backfill height.

3.2 5.5-Inch- (140-mm-) Thick Flat ICF Basement Wall

A flat ICF foundation wall is selected from Table 3.4 of the *Prescriptive Method* for foundation walls constructed in soil with an equivalent fluid density of 30 pcf (481 kg/m^3). The foundation wall is 9 feet (1.2 m) high and has 6 feet (1.8 m) of unbalanced backfill. Table 3.4 shows that the foundation wall is required to have one No. 3 bar at 10 inches (254 mm) on center for minimum vertical wall reinforcement. Calculate the capacity and check the adequacy of the 5.5-inch- (140-mm-) thick flat ICF foundation wall.

Using the values in Figure B1.2 and the material properties in Section 2.0, compute the amount of vertical wall reinforcement required in Construction Cases A1 through A4.

3.2.1 Construction Case A1

Figures B3.10 and B3.11 summarize the nominal and factored loads calculated in this section. Refer to Section 7.0 for typical loading condition equations. The wall moments are calculated by multiplying the axial load by the assumed eccentricities of the roof, wall above, and floor as applicable.

Dead Loads

First Floor	$0.5(32\ ft)(10\ psf) = 160\ plf$	[2.3 *kN/m*]
Roof and Ceiling	$0.5(40\ ft)(12\ psf) = 240\ plf$	[3.5 *kN/m*]
First-Story Wood Wall	$(8.0\ ft)(6\ psf) = 48\ plf$ @ *base of wall*	[0.7 *kN/m*]
Foundation ICF Wall	$0.5(9.0\ ft)(68.75\ psf) = 309\ plf$ @ *mid-height*	[4.5 *kN/m*]

Live Loads

First Floor	$0.5(32\ ft)(40\ psf) = 640\ plf$	[9.3 *kN/m*]
Roof and Ceiling	$0.5(40\ ft)(0\ psf) = 0\ plf$	[0 *kN/m*]
Attic	$0.5(40\ ft)(0\ psf) = 0\ plf$	[0 *kN/m*]

Foundation Wall Moments

Dead Load $_{@top}$ $\quad [240\ plf + 48\ plf + 160\ plf](0\ in) = 0\ in-lb/lf$ [0 *N-m/m*]

Live Load $_{@top}$ $\quad (0\ plf + 0\ plf + 640\ plf)(0\ in) = 0\ in-lb/lf$ [0 *N-m/m*]

$$Earth\ Load_{midht} \quad -\frac{(30\ pcf)(6\ ft)(4.5\ ft)^2}{2} + \frac{(30\ pcf)(4.5\ ft)^3}{6} + 420\ plf(4.5\ ft) = 523\ ft-lb/ft$$
[2.3 *kN-m/m*]

$$Earth\ Load_{@x} \quad -\frac{(30\ pcf)(6\ ft)(3.17\ ft)^2}{2} + \frac{(30\ pcf)(3.17\ ft)^3}{6} + 420\ plf(3.17\ ft) = 586\ ft-lb/ft$$
[2.6 *kN-m/m*]

Parallel Shear

The unbalanced backfill height is greater than 4 feet (1.2 m). Assume all basement walls in the building have 6 feet (1.8 m) of unbalanced backfill height. Parallel shear is neglected since the foundation walls are restrained by the soil lateral pressure on all sides.

Perpendicular Shear

Refer to Section 7.0 for variable definitions.

$$V_{top} = \frac{ql^3}{6L} = \frac{(30\ pcf)(6ft)^3}{6(9ft)} = 120\ plf \qquad [1.8\ kN/m]$$

$$V_{midht} = qlx - \frac{qx^2}{2} - V_{bottom} = (30\ pcf)(6\ ft)(4.5\ ft) - \frac{(30\ pcf)(4.5\ ft)^2}{2} - 420\ plf = 86\ plf \qquad [1.3\ kN/m]$$

$$V_{bottom} = \frac{ql^2}{2} - V_{top} = \frac{(30\ pcf)(6\ ft)^2}{2} - 120\ plf = 420\ plf \qquad [6.1\ kN/m]$$

NOMINAL STRUCTURAL LOAD SUMMARY

Story	Vertical Location within Wall	Dead		Live		Wind			Earth		
		Axial (plf)	Moment (in-lb/lf)	Axial (plf)	Moment (in-lb/lf)	Moment (in-lb/lf)	$Shear_{,perp}$ (plf)	$Shear_{,par}$ (lb)	Moment (in-lb/lf)	$Shear_{,perp}$ (plf)	$Shear_{,par}$ (lb)
Foundation	Top	448	0	640	0				0	120	
	Midheight	757	0	640	0				6,278	86	
	x = 3.17 ft	849	0	640	0				7,035	0	
	Bottom	1,067	0	640	0				0	420	N/A

For SI: 1 plf = 14.59390 N/m; 1 in-lb/lf = 0.3706850 N-m/m; 1 lb = 4.448222 N

Figure B3.10 Nominal Load Summary for Construction Case A1

Figure B3.10 is a summary sheet of the loading conditions described above. To determine the moments at various locations in each wall story, refer to Section 7.0. Figure B3.11 is a summary sheet of the ACI [B3] factored loads for each ICF wall story. The values listed in Figure B3.11 are determined by substituting the values from Figure B3.10 into the equations listed in the left column of Figure B3.11.

FACTORED STRUCTURAL LOAD SUMMARY

FOUNDATION WALL

ACI 318 Load Combinations[1]	Vertical Location within Wall	Factored Loads										
		Dead		Live		Earth			Total			
		Axial (plf)	Moment (in-lb/lf)	Axial (plf)	Moment (in-lb/lf)	Moment (in-lb/lf)	$Shear_{,perp}$ (plf)	$Shear_{,par}$ (lb)	Axial (plf)	Moment (in-lb/lf)	$Shear_{,perp}$ (plf)	$Shear_{,par}$ (lb)
1. U = 1.4D + 1.7L	Top	627	0	1,088	0				1,715	0		
	Midheight	1,060	0	1,088	0				2,148	0		
	x = 3.17 ft	1,188	0	1,088	0				2,276	0		
	Bottom	1,493	0	1,088	0				2,581	0		
2. U = 1.4D + 1.7L + 1.7H	Top	627	0	1,088	0	0	204	N/A	1,715	0	204	N/A
	Midheight	1,060	0	1,088	0	10,673	147		2,148	10,673	147	
	x = 3.17 ft	1,188	0	1,088	0	11,960	0		2,276	11,960	0	
	Bottom	1,493	0	1,088	0	0	714		2,581	0	714	
3. U = 0.9D + 1.7H	Top	403	0			0	204	N/A	403	0	204	N/A
	Midheight	682	0			10,673	147		682	10,673	147	
	x = 3.17 ft	764	0			11,960	0		764	11,960	0	
	Bottom	960	0			0	714		960	0	714	

[1] D = Dead Load, L = Live Load, W = Wind Load, H = Earth Load

For SI: 1 plf = 14.59390 N/m; 1 in-lb/lf = 0.3706850 N-m/m; 1 lb = 4.448222 N

Figure B3.11 Factored Load Summary for Construction Case A1

Assume the section does not require reinforcement.

Check Perpendicular Shear

According to Figure B3.11, the critical factored perpendicular shear load, V_u, experienced by the foundation wall occurs at the bottom of the wall story due to ACI [B3] Load Combinations (2) and (3).

$$V_u = 714\ plf \qquad [10.4\ kN/m]$$

$$\phi V_n = 0.65\left(\frac{4}{3}\right)\sqrt{f'_c}\,bh = 0.65\left(\frac{4}{3}\right)\sqrt{2{,}500\,psi}\,(12\,in)(5.5\,in) = 2{,}860\ plf \qquad [41.7\ kN/m]$$

$$V_u \le \phi V_n \quad OK$$

Check Compression and Tension

According to Figure B3.11, the critical maximum total moment, M_u, experienced by the foundation story wall occurs at x = 3.17 feet (0.97 m) due to ACI [B3] Load Combinations (2) and (3). The corresponding total factored axial load, P_u, is also taken at x = 3.17 feet (0.97 m) of the foundation wall based on ACI [B3] Load Combinations (2) and (3). The minimum factored moment required for design is specified in ACI 22.6.3 [B3].

$$M_u = 11{,}960\,in-lb/ft = 1.0\,ft-kip/lf \longleftarrow GOVERNS \qquad [4.4\ kN\text{-}m/m]$$

$$P_u = 2{,}276\,plf = 2.3\,kip/lf \ \ (ACI\ LOAD\ COMBINATION\ 2) \qquad [33.2\ kN/m]$$

$$P_u = 764\,plf = 0.76\,kip/lf \ \ (ACI\ LOAD\ COMBINATION\ 3) \qquad [11\ kN/m]$$

$$M_{u,min} = 0.1hP_u = 0.1(5.5\,in)(2{,}276\,plf) = 1{,}252\,in-lb/lf$$
$$= 0.10\,ft-kip/lf\ (ACI\ LOAD\ COMBINATION\ 2) \qquad [0.46\ kN\text{-}m/m]$$

$$M_{u,min} = 0.1hP_u = 0.1(5.5\,in)(764\,plf) = 420\,in-lb/lf$$
$$= 0.04\,ft-kip/lf\ \ (ACI\ LOAD\ COMBINATION\ 3) \qquad [0.16\ kN\text{-}m/m]$$

Plot M_u and P_u on the interaction diagram for a 5.5-inch- (140-mm-) thick flat wall shown in Figure B3.12. The plotted point, "A1", lies below the lower tension boundary for ACI [B3] Load Combination (3); therefore, a 5.5-inch- (140-mm-) thick flat wall structural plain concrete wall is insufficient for the given loading conditions.

Design a structurally reinforced wall section with No. 3 bar at 12 inches (305 mm) on center.

Check Perpendicular Shear

According to Figure B3.10, the critical factored perpendicular shear load, V_u, experienced by the foundation wall occurs at the bottom of the wall story due to ACI [B3] Load Combinations (2) and (3).

$$V_u = 714\ plf \qquad [10.4\ kN/m]$$

$$\phi V_n = 0.85(2)\sqrt{f'_c}\,b_w d = 0.85(2)\sqrt{2{,}500\,psi}\,(12\,in)\left(\frac{5.5}{2}in\right) = 2{,}805\ plf \qquad [40.9\ kN/m]$$

$$V_u \leq \phi V_n \quad OK$$

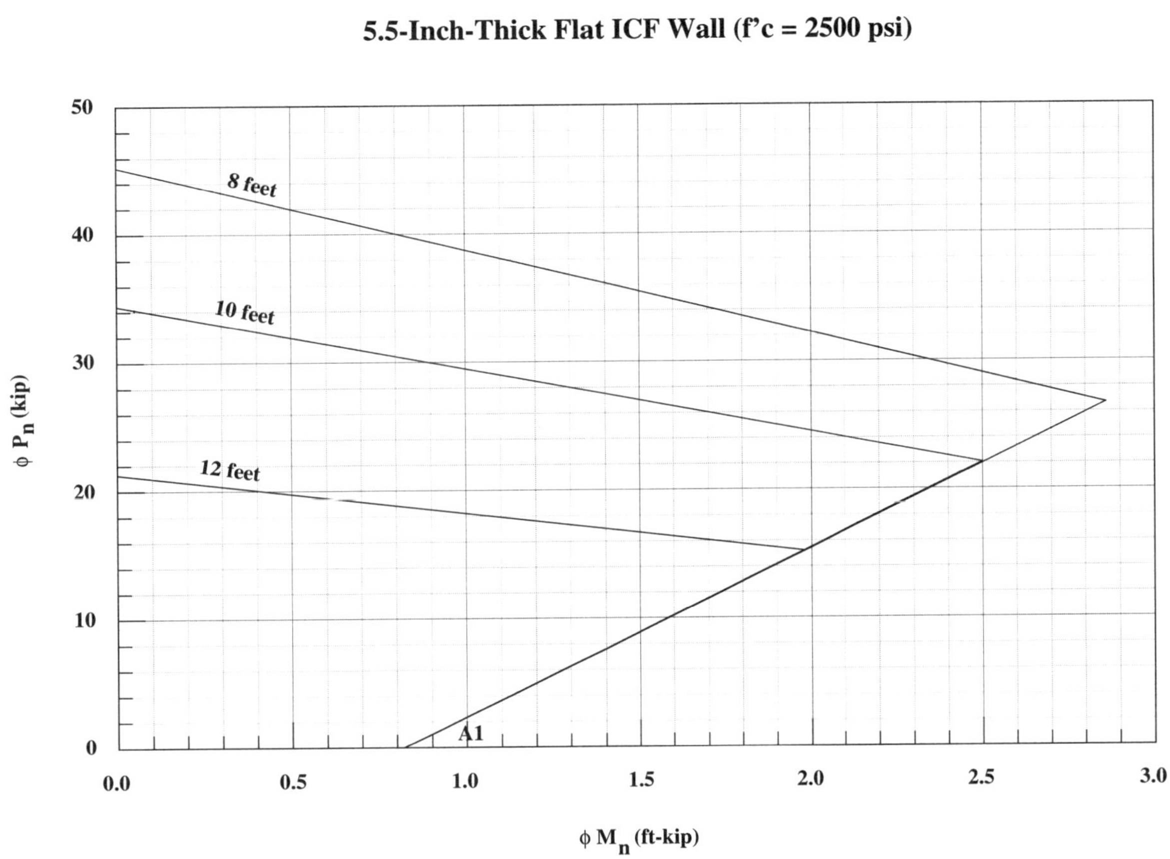

Figure B3.12 Structural Plain Concrete Wall Interaction Diagram

Determine Sway and Slenderness

Most homes built with ICF wall systems can be categorized as nonsway frames in accordance with ACI [B3]. To determine whether the wall is slender, the radius of gyration is approximated as *0.3h* for rectangular compression members, *k* is assumed to be 1.0 since the wall is tied to the footing below and the floor above, and M_1 is assumed to be 0.

$$\frac{kl_u}{r} < 34 - 12\frac{M_1}{M_2}$$

$$\frac{(1)(9\,ft)\left(\frac{12\,in}{ft}\right)}{0.3(5.5\,in)} = 65.45$$

$65.45 \leq 34$ *Slender*

ACI 10.10.2 [B3] allows the moment magnifier method to be used for walls with a slenderness ratio, kl_u/r, of 100 or less to account for slenderness effects in a wall.

Determine Magnified Moment

With one exception, the equations below are taken from ACI 10.12 [B3]. The equation for *EI*, as listed in ACI 10.12.3 [B3], is applicable to wall sections that contain a double layer of reinforcement. Given that ICFs contain only one layer of reinforcement, the equation for *EI* noted below is used instead [B7].

According to Figure B3.11, the critical maximum total moment, M_u, experienced by the foundation story wall occurs at x = 3.17 feet (0.97 m) due to ACI [B3] Load Combination (3). The corresponding total factored axial load, P_u, is also taken at x = 3.17 feet (0.97 m) of the foundation wall based on ACI [B3] Load Combination (3).

$$M_u = 11{,}960\,in-lb/lf\,(1\,ft) = 11{,}960\,in-lb \qquad [1.4\ kN\text{-}m]$$

$$P_u = 764\,plf\,(1\,ft) = 764\,lb \qquad [3.4\ kN]$$

$$E_c = 57{,}000\sqrt{f_c'} = 57{,}000\sqrt{2{,}500\,psi} = 2{,}850{,}000\,psi \qquad [19.7\ GPa]$$

$$\beta_d = \frac{P_{u,dead}}{P_u} = \frac{764\,lb}{764\,lb} = 1.0$$

$$\rho = \frac{A_s}{A_g} = \frac{0.110\,in^2}{(5.5\,in)(12\,in)} = 0.0017$$

$$\beta = 0.9 + 0.5\beta_d^2 - 12\rho \geq 1.0 = 0.9 + 0.5(1.0)^2 - 12(0.0017) = 1.38$$

$$e = \frac{M_u}{P_u} = \frac{11{,}960\,in-lb}{764\,lb} = 15.65\,in \qquad [398\ mm]$$

$$EI = \frac{0.4E_cI_g}{\beta} \geq \frac{E_cI_g\left(0.5 - {}^{e}\!/_{h}\right)}{\beta} \geq \frac{0.1E_cI_g}{\beta}$$

$$= \frac{0.1(2{,}850{,}000\,psi)\,{}^{(12\,in)(5.5\,in)^3}\!/_{12}}{1.38} = 34{,}360{,}054\,psi \qquad [237\ GPa]$$

$$M_{2,min} = P_u(0.6 + 0.03h) = 764\,lb(0.6 + 0.03(5.5\,in)) = 585\,in-lb \qquad [66.1\ kN\text{-}m]$$

$$C_m = 1.0\ \textit{for members with transverse loads between supports}$$

$$P_c = \frac{\pi^2\,EI}{(kl_u)^2} = \frac{\pi^2(34{,}360{,}054\,psi)}{(1.0)(12\,in/ft)\,(9\,ft)} = 29{,}074\,lb \qquad [129\ kN]$$

$$\delta_{ns} = \frac{C_m}{1-\left({}^{P_u}\!/_{0.75P_c}\right)} \geq 1.0 = \frac{1.0}{1-\left({}^{764\,lb}\!/_{0.75(29{,}074\,lb)}\right)} = 1.036$$

$$M_{ns} = \delta M_u = 1.036(11{,}960\,in-lb) = 12{,}391\,in-lb = 1.032\,ft-kip \qquad [1.4\ kN\text{-}m]$$

Determine Reinforcement

To determine if the wall section is adequately reinforced, plot the magnified moment and the corresponding total factored axial load on an interaction diagram shown in Figure B3.13. The plotted point, "A", lies below the interaction diagram; therefore, a No. 3 bar at 12 inches (305 mm) on center is not adequate.

Design a structurally reinforced wall section with one No. 3 bar at 10 inches (254 mm) on center.

Check Perpendicular Shear

According to Figure B3.11, the critical factored perpendicular shear load, V_u, experienced by the foundation wall occurs at the bottom of the wall story due to ACI [B3] Load Combinations (2) and (3).

$$V_u = 714 \ plf \qquad [10.4 \ kN/m]$$

$$\phi V_n = 0.85(2)\sqrt{f'_c}\, b_w d = 0.85(2)\sqrt{2{,}500 psi}\,(10 in)\left(\frac{5.5 in}{2}\right) = 2{,}338 \ plf \qquad [34.1 \ kN/m]$$

$$V_u \leq \phi V_n \quad OK$$

Determine Magnified Moment

With one exception, the equations below are taken from ACI 10.12 [B3]. The equation for *EI*, as listed in ACI 10.12.3 [B3], is applicable to wall sections that contain a double layer of reinforcement. Given that ICFs contain only one layer of reinforcement, the equation for *EI* noted below is used instead [B7].

According to Figure B3.11, the critical maximum total moment, M_u, experienced by the foundation story wall occurs at x = 3.17 feet (0.97 m) due to ACI [B3] Load Combination (3). The corresponding total factored axial load, P_u, is also taken at x = 3.17 feet (0.97 m) of the foundation wall based on ACI [B3] Load Combination (3).

$$M_2 = 11{,}960 in - lb / lf\,(0.83 ft) = 9{,}927 in - lb \qquad [1.1 \ kN\text{-}m]$$

$$P_u = 764 lb / lf\,(0.83 ft) = 634 lb \qquad [2.8 \ kN]$$

$$E_c = 57{,}000\sqrt{f_c'} = 57{,}000\sqrt{2{,}500 psi} = 2{,}850{,}000 psi \qquad [19.7 \ GPa]$$

$$\beta_d = \frac{P_{u,dead}}{P_u} = \frac{634 lb}{634 lb} = 1.0$$

$$\rho = \frac{A_s}{A_g} = \frac{0.110 in^2}{(5.5 in)(12 in)} = 0.0017$$

$$\beta = 0.9 + 0.5\beta_d^2 - 12\rho \geq 1.0 = 0.9 + 0.5(1.0)^2 - 12(0.0017) = 1.38$$

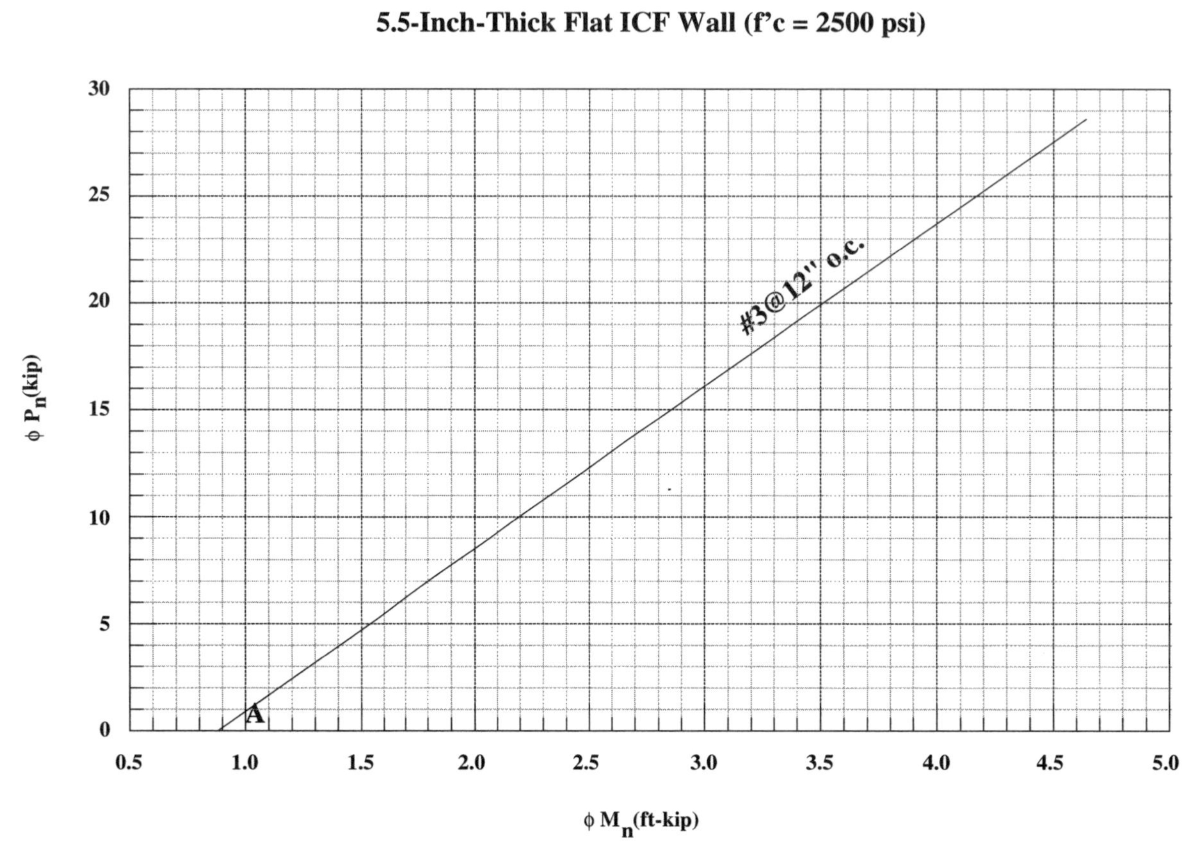

Figure B3.13 Structural Reinforced 5.5-Inch- (140-mm-) Thick Flat Concrete Wall Interaction Diagram

$$EI = \frac{0.4E_c I_g}{\beta} \geq \frac{E_c I_g \left(0.5 - {}^{e}\!/_{h}\right)}{\beta} \geq \frac{0.1E_c I_g}{\beta}$$

$$= \frac{0.1(2{,}850{,}000\,psi)\left(12\,in\right)\left(5.5\,in\right)^3 / 12}{1.38} = 34{,}360{,}054\,psi$$ [237 *GPa*]

$$M_{2,min} = P_u(0.6 + 0.03h) = 634\,lb(0.6 + 0.03(5.5\,in)) = 485\,in - lb$$ [55 *N-m*]

$$C_m = 1.0 \textit{ for members with transverse loads between supports}$$

$$P_c = \frac{\pi^2 EI}{\left(kl_u\right)^2} = \frac{\pi^2 \left(34{,}360{,}054\,psi\right)}{(1.0)(12\,in/ft)\,(9\,ft)} = 29{,}074\,lb$$ [129 *kN*]

$$\delta_{ns} = \frac{C_m}{1 - \left({}^{P_u}\!/_{0.75P_c}\right)} \geq 1.0 = \frac{1.0}{1 - \left({}^{634\,lb}\!/_{0.75(29{,}074\,lb)}\right)} = 1.03$$

$$M_{ns} = \delta M_u = 1.03(9{,}927\,in - lb) = 10{,}225\,in - lb = 0.85\,ft - kip$$ [1.2 *kN-m*]

Determine Reinforcement

To determine if the wall section is adequately reinforced, plot the magnified moment and the corresponding total factored axial load on an interaction diagram shown in Figure B3.14. The plotted point, "B", lies above the plotted line in the interaction diagram; therefore, one vertical No. 3 bar at 10 inches (254 mm) on center is adequate.

Tests [B6] have shown that horizontal wall reinforcement spacing limited to 48 inches (1.2 m) on center results in reliable performance; therefore, install one No. 4 bar at 48 inches (1.2 m) on center horizontally. At least one continuous horizontal reinforcement bar should be placed within the top 12 inches (305 mm) of the wall story.

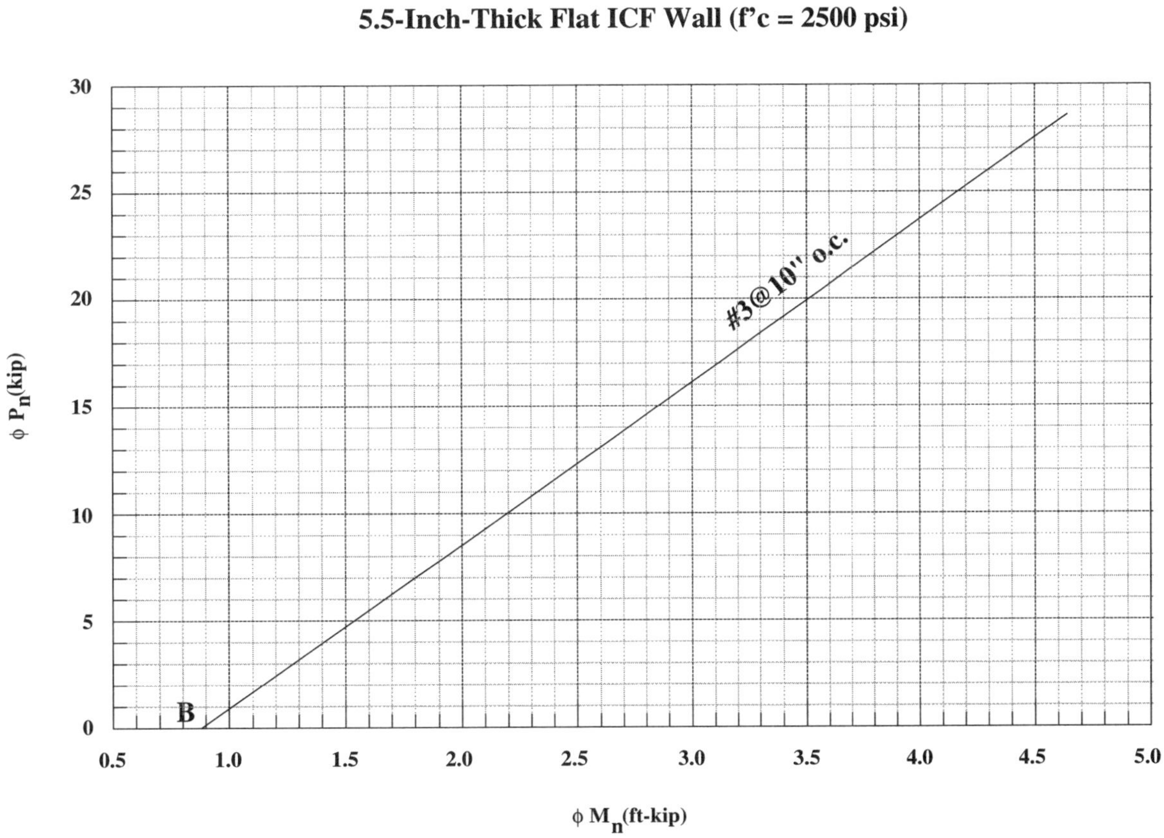

Figure B3.14 Structural Reinforced 5.5-Inch- (140-mm-) Thick Flat Concrete Wall Interaction Diagram

Check Deflection

For below-grade walls, a conservative deflection limit of *L/240* for service live loads is used because earth loads are immediate and are not expected to change significantly with time. To calculate wall deflection at service load levels, effective section properties of the assumed cracked concrete section must be established. According to test results [B6], calculating deflection by using $0.1E_cI_g$ was found to be conservative but more accurate than calculating deflection based on the cracking moment per ACI 9.5.2.3 [B3]. When using the moment magnifier concept, it is recommended that the calculated moment magnification factor be applied to the service load moments to be used in conducting the deflection calculations. Assume a gypsum board interior finish exposed to view. For simplicity, when calculating the maximum deflection, assume the earth load acts on the entire wall height.

$$\Delta_{maximum} = \frac{0.01304(\delta)(0.5)ql^5}{0.1E_cI_g} = \frac{0.01304(1.03)(0.5)(30\,pcf)(1\,ft)(9\,ft)^5}{(0.1)(2{,}850{,}000\,psi)\left(\frac{(5.5\,in)(12\,in)^3}{12}\right)}\left(\frac{1{,}728\,in^3}{ft^3}\right) = 0.091\,in \qquad [2.3\,mm]$$

$$\Delta_{allowable} = \frac{L}{240} = \frac{(9\,ft)\left(\frac{12\,in}{ft}\right)}{240} = 0.45\,in \qquad [11.4\,mm]$$

$$\Delta_{actual} \leq \Delta_{allowable} \qquad OK$$

3.2.2 Construction Case A2

Figures B3.15 and B3.16 summarize the nominal and factored loads calculated in this section. Refer to Section 7.0 for typical loading condition equations. The wall moments are calculated by multiplying the axial load by the assumed eccentricities of the roof, wall above, and floor as applicable.

Dead Loads

First Floor	$0.5(32\,ft)(15\,psf) = 240\,plf$	[3.5 *kN/m*]
Second Floor	$0.5(32\,ft)(15\,psf) = 240\,plf$	[3.5 *kN/m*]
Roof and Ceiling	$0.5(40\,ft)(16\,psf) = 320\,plf$	[4.7 *kN/m*]
First-Story Wood Wall	$(10\,ft)(11\,psf) = 110\,plf$ @ *base of wall*	[1.6 *kN/m*]
Second-Story Wood Wall	$(10\,ft)(11\,psf) = 110\,plf$ @ *base of wall*	[1.6 *kN/m*]
Foundation ICF Wall	$0.5(9.0\,ft)(68.75\,psf) = 309\,plf$ @ *mid-height*	[4.5 *kN/m*]

Live Loads

First Floor	$0.5(32\,ft)(40\,psf) = 640\,plf$	[9.3 *kN/m*]
Second Floor	$0.5(32\,ft)(30\,psf) = 480\,plf$	[7.0 *kN/m*]
Roof and Ceiling	$0.5(40\,ft)(49\,psf) = 980\,plf$	[14.3 *kN/m*]
Attic	$0.5(40\,ft)(20\,psf) = 400\,plf$	[5.8 *kN/m*]

Foundation Wall Moments

Dead Load $_{@top}$ $\left[2(240)plf + 2(110)plf + 320\ plf\right](0\,in) = 0\,in - lb/lf$ [0 *N-m/m*]

Live Load $_{@top}$ $\left[(640\ plf + 480\ plf + 400\ plf + 980\ plf)\right](0\,in) = 0\,in - lb/lf$ [0 *N-m/m*]

Earth Load *Unchanged from Construction Case A*1*; refer to Section* 3.2.1.

Parallel Shear

*Unchanged from Construction Case A*1*; refer to Section* 3.2.1.

Perpendicular Shear

*Unchanged from Construction Case A*1*; refer to Section* 3.2.1.

NOMINAL STRUCTURAL LOAD SUMMARY											
Story	Vertical Location within Wall	Dead		Live		Wind			Earth		
		Axial (plf)	Moment (in-lb/lf)	Axial (plf)	Moment (in-lb/lf)	Moment (in-lb/lf)	Shear,perp (plf)	Shear,par (lb)	Moment (in-lb/lf)	Shear,perp (plf)	Shear,par (lb)
Foundation	Top	1,020	0	2,500	0				0	120	
	Midheight	1,329	0	2,500	0				6,278	86	
	x = 3.17 ft	1,443	0	2,500	0				7,035	0	
	Bottom	1,639	0	2,500	0				0	420	N/A

For SI: 1 plf = 14.59390 N/m; 1 in-lb/lf = 0.3706850 N-m/m; 1 lb = 4.448222 N

Figure B3.15 Nominal Load Summary for Construction Case A2

Figure B3.15 is a summary sheet of the loading conditions described above. To determine the moments at various locations in each wall story, refer to Section 7.0. Figure B3.16 is a summary sheet of the ACI [B3] factored loads for each ICF wall story. The values listed in Figure B3.16 are determined by substituting the values from Figure B3.15 into the equations listed in the left column of Figure B3.16.

FACTORED STRUCTURAL LOAD SUMMARY												
FOUNDATION WALL												
ACI 318 Load Combinations[1]	Vertical Location within Wall	Factored Loads										
		Dead		Live		Earth			Total			
		Axial (plf)	Moment (in-lb/lf)	Axial (plf)	Moment (in-lb/lf)	Moment (in-lb/lf)	Shear,perp (plf)	Shear,par (lb)	Axial (plf)	Moment (in-lb/lf)	Shear,perp (plf)	Shear,par (lb)
1. U = 1.4D + 1.7L	Top	1,428	0	4,250	0				5,678	0		
	Midheight	1,861	0	4,250	0				6,111	0		
	x = 3.17 ft	2,020	0	4,250	0				6,270	0		
	Bottom	2,295	0	4,250	0				6,545	0		
2. U = 1.4D + 1.7L + 1.7H	Top	1,428	0	4,250	0	0	204	N/A	5,678	0	204	N/A
	Midheight	1,861	0	4,250	0	10,673	146		6,111	10,673	146	
	x = 3.17 ft	2,020	0	4,250	0	11,960	0		6,270	11,960	0	
	Bottom	2,295	0	4,250	0	0	714		6,545	0	714	
3. U = 0.9D + 1.7H	Top	918	0			0	204	N/A	918	0	204	N/A
	Midheight	1,196	0			10,673	146		1,196	10,673	146	
	x = 3.17 ft	1,299	0			11,960	0		1,299	11,960	0	
	Bottom	1,475	0			0	714		1,475	0	714	

[1] D = Dead Load, L = Live Load, W = Wind Load, H = Earth Load

For SI: 1 plf = 14.59390 N/m; 1 in-lb/lf = 0.3706850 N-m/m; 1 lb = 4.448222 N

Figure B3.16 Factored Load Summary for Construction Case A2

Assume the section does not require reinforcement.

Check Perpendicular Shear

According to Figure B3.16, the critical factored perpendicular shear load, V_u, is the same as that in Construction Case A1; refer to Section 3.2.1 for calculations for a structural plain concrete section.

Check Compression and Tension

According to Figure B3.16, the critical maximum total moment, M_u, experienced by the foundation story wall occurs at x = 3.17 feet (0.97 m) due to ACI [B3] Load Combinations (2) and (3). The corresponding total factored axial load, P_u, is also taken at x = 3.17 feet (0.97 m) of the foundation wall based on ACI [B3] Load Combinations (2) and (3). The minimum factored moment required for design is specified in ACI 22.6.3 [B3].

$$M_u = 11{,}960\, in - lb / lf = 1.0\, ft - kip / lf \longleftarrow GOVERNS \quad [4.4\ kN\text{-}m/m]$$
$$P_u = 6{,}270\, plf = 6.3\, kip / lf \;\; (ACI\ LOAD\ COMBINATION\ 2) \quad [91.5\ kN/m]$$
$$P_u = 1{,}299\, plf = 1.3\, kip / lf \;\; (ACI\ LOAD\ COMBINATION\ 3) \quad [18\ kN/m]$$
$$M_{u,min} = 0.1hP_u = 0.1(5.5\, in)(6{,}270\, plf) = 3{,}449\, in - lb / lf = 0.29\, ft - kip / lf \;\; (ACI\ LOAD\ COMBINATION\ 2) \quad [1.3\ kN\text{-}m/m]$$
$$M_{u,min} = 0.1hP_u = 0.1(5.5\, in)(1{,}299\, plf) = 715\, in - lb / lf = 0.06\, ft - kip / lf \;\; (ACI\ LOAD\ COMBINATION\ 3) \quad [0.27\ kN\text{-}m/m]$$

Plot M_u and P_u on the interaction diagram for a 5.5-inch- (140-mm-) thick flat wall shown in Figure B3.12. The plotted point, "A1", lies below the lower tension boundary for ACI [B3] Load Combination (3); therefore, a 5.5-inch- (140-mm-) thick flat wall structural plain concrete wall is insufficient for the given loading conditions.

Design a structurally reinforced wall section with No. 3 bar at 12 inches (305 mm) on center.

Check Perpendicular Shear

Unchanged from Construction Case A1; refer to Section 3.2.1 for calculations for a structural reinforced section, one No. 3 bar at 12 inches (305 mm) on center.

Determine Sway and Slenderness

Unchanged from Construction Case A1; refer to Section 3.2.1 for calculations for a structural reinforced section, one No. 3 bar at 12 inches (305 mm) on center.

Determine Magnified Moment

With one exception, the equations below are taken from ACI 10.12 [B3]. The equation for *EI*, as listed in ACI 10.12.3 [B3], is applicable to wall sections that contain a double layer of reinforcement. Given that ICFs contain only one layer of reinforcement, the equation for *EI* noted below is used instead [B7].

According to Figure B3.16, the critical maximum total moment, M_u, experienced by the foundation story wall occurs at x = 3.17 feet (0.97 m) due to ACI [B3] Load Combination (3). The corresponding total factored axial load, P_u, is also taken at x = 3.17 feet (0.97 m) of the foundation wall based on ACI [B3] Load Combination (3).

$$M_u = 11{,}960\, in-lb/lf\,(1\,ft) = 11{,}960\, in-lb \qquad [1.4\ kN\text{-}m]$$

$$P_u = 1{,}299\ plf\,(1\,ft) = 1{,}299\ lb \qquad [5.8\ kN]$$

$$E_c = 57{,}000\sqrt{f_c'} = 57{,}000\sqrt{2{,}500\ psi} = 2{,}850{,}000\ psi \qquad [19.7\ GPa]$$

$$\beta_d = \frac{P_{u,dead}}{P_u} = \frac{1{,}299\ lb}{1{,}299\ lb} = 1.0$$

$$\rho = \frac{A_s}{A_g} = \frac{0.110\, in^2}{(5.5\, in)(12\, in)} = 0.0017$$

$$\beta = 0.9 + 0.5\beta_d^{\,2} - 12\rho \geq 1.0 = 0.9 + 0.5(1.0)^2 - 12(0.0017) = 1.38$$

$$e = \frac{M_u}{P_u} = \frac{11{,}960\, in-lb}{1{,}299\ lb} = 9.21\ in \qquad [234\ mm]$$

$$EI = \frac{0.4E_cI_g}{\beta} \geq \frac{E_cI_g\left(0.5 - \frac{e}{h}\right)}{\beta} \geq \frac{0.1E_cI_g}{\beta} = \frac{0.1(2{,}850{,}000\ psi)\frac{(12\, in)(5.5\, in)^3}{12}}{1.38} \qquad [237\ GPa]$$

$$= 34{,}360{,}054\ psi$$

$$M_{2,min} = P_u(0.6 + 0.03h) = 1{,}299\ lb(0.6 + 0.03(5.5\, in)) = 994\ in-lb \qquad [112\ N\text{-}m]$$

$$C_m = 1.0\ \textit{for members with transverse loads between supports}$$

$$P_c = \frac{\pi^2\, EI}{(kl_u)^2} = \frac{\pi^2\,(34{,}360{,}054\ psi)}{(1.0)(12\, in/ft)^2(9\,ft)^2} = 29{,}074\ lb \qquad [129\ kN]$$

$$\delta_{ns} = \frac{C_m}{1 - \left(\frac{P_u}{0.75P_c}\right)} \geq 1.0 = \frac{1.0}{1 - \left(\frac{1{,}299\ lb}{0.75(29{,}074\ lb)}\right)} = 1.063$$

$$M_{ns} = \delta M_u = 1.063(11{,}960\, in-lb) = 12{,}713\, in-lb = 1.06\ ft-kip \qquad [1.4\ kN\text{-}m]$$

Determine Reinforcement

To determine if the wall section is adequately reinforced, plot the magnified moment and the corresponding total factored axial load on an interaction diagram shown in Figure B3.17. The plotted point, "C", lies above the plotted line of the interaction diagram; therefore, one No. 3 bar at 12 inches (305 mm) on center is adequate.

Tests [B6] have shown that horizontal wall reinforcement spacing limited to 48 inches (1.2 m) on center results in reliable performance; therefore, install one No. 4 bar at 48 inches (1.2 m) on center horizontally. At least one continuous horizontal reinforcement bar should be placed within the top 12 inches (305 mm) of the wall story.

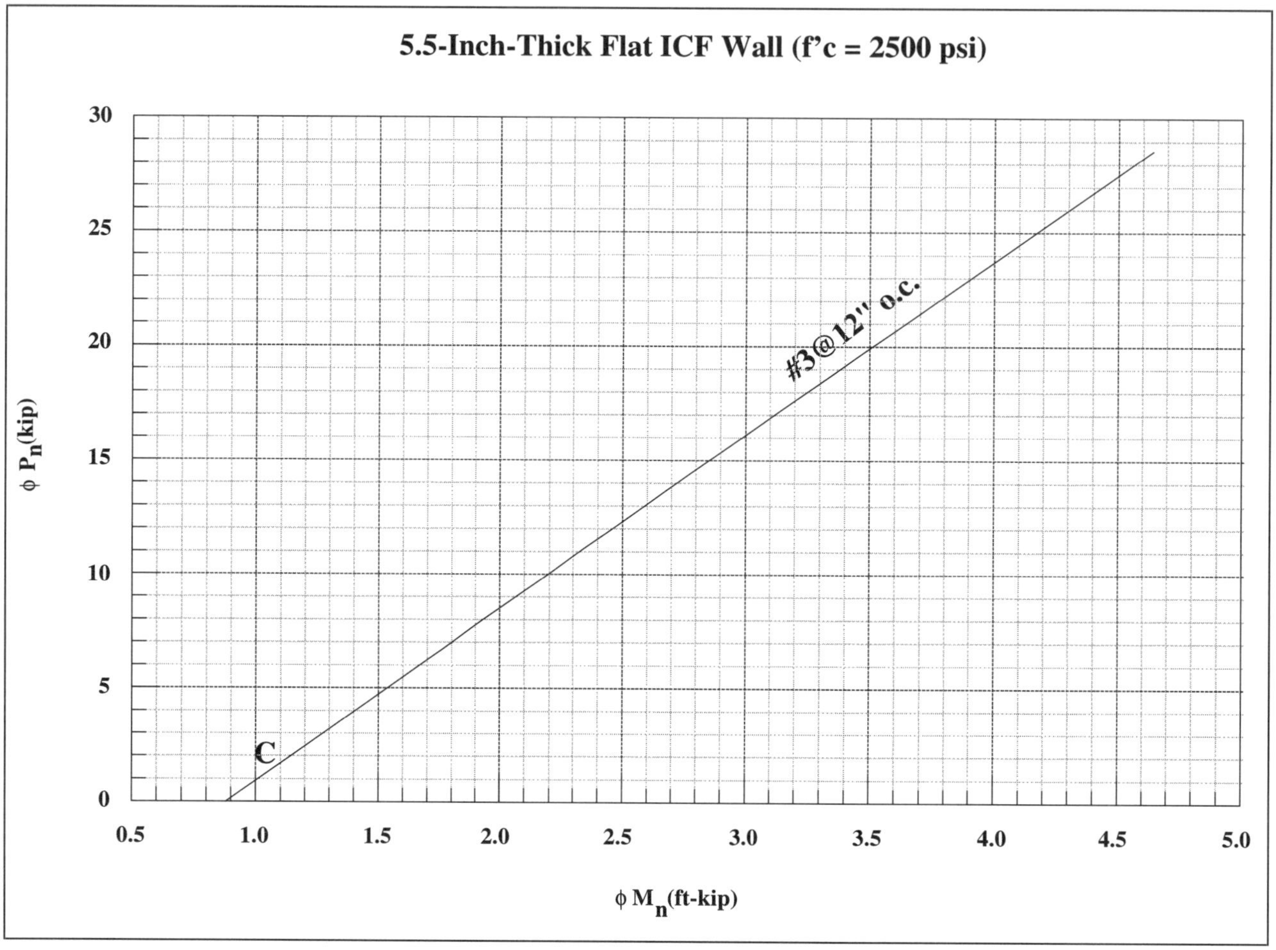

Figure B3.17 Structural Reinforced 5.5-Inch- (140-mm-) Thick Flat Concrete Wall Interaction Diagram

<u>Check Deflection</u>

For below-grade walls, a conservative deflection limit of *L/240* for service live loads is used because earth loads are immediate and are not expected to change significantly with time. To calculate wall deflection at service load levels, effective section properties of the assumed cracked concrete section must be established. According to test results [B6], calculating deflection by using $0.1E_cI_g$ was found to be conservative but more accurate than calculating deflection based on the cracking moment per ACI 9.5.2.3 [B3]. When using the moment magnifier concept, it is recommended that the calculated moment magnification factor be applied to the service load moments to be used in conducting the deflection calculations. Assume a gypsum board interior finish exposed to view. For simplicity, when calculating the maximum deflection, assume the earth load acts on the entire wall height.

$$\Delta_{maximum} = \frac{0.01304(\delta)(0.5)ql^5}{0.1E_cI_g} = \frac{0.01304(1.06)(0.5)(30\,pcf)(1\,ft)(9\,ft)^5}{(0.1)(2{,}850{,}000\,psi)\left(\frac{(5.5\,in)(12\,in)^3}{12}\right)}\left(\frac{1{,}728\,in^3}{ft^3}\right) = 0.094\,in \qquad [2.4\,mm]$$

$$\Delta_{allowable} = \frac{L}{240} = \frac{(9\,ft)\left(\frac{12\,in}{ft}\right)}{240} = 0.45\,in \qquad [11.4\,mm]$$

$\Delta_{actual} \le \Delta_{allowable}$ *OK*

3.2.3 Construction Case A3

Figures B3.18 and B3.19 summarize the nominal and factored loads calculated in this section. Refer to Section 7.0 for typical loading condition equations. The wall moments are calculated by multiplying the axial load by the assumed eccentricities of the roof, wall above, and floor as applicable.

<u>*Dead Loads*</u>

First Floor	$0.5(32\,ft)(10\,psf) = 160\,plf$	[2.3 *kN/m*]
Roof and Ceiling	$0.5(40\,ft)(12\,psf) = 240\,plf$	[3.5 *kN/m*]
First-Story ICF Wall	$0.5(8\,ft)(50\,psf) = 200\,plf$ *@ mid-height*	[2.9 *kN/m*]
Foundation ICF Wall	$0.5(9.0\,ft)(68.75\,psf) = 309\,plf$ *@ mid-height*	[4.5 *kN/m*]

<u>*Live Loads*</u>

First Floor	$0.5(32\,ft)(10\,psf) = 160\,plf$	[2.3 *kN/m*]
Roof and Ceiling	$0.5(40\,ft)(0\,psf) = 0\,plf$	[0 *kN/m*]
Attic	$0.5(40\,ft)(0\,psf) = 0\,plf$	[0 *kN/m*]

<u>*Foundation Wall Moments*</u>

Dead Load $_{@top}$	$[240\,plf + 2(200\,plf)](0.75\,in) + (160\,plf)(-2.75\,in) = 40\,in-lb/lf$	[14.8 *N-m/m*]
Live Load $_{@top}$	$(0\,plf + 0\,plf)(0.75\,in) + (160\,plf)(-2.75\,in) = -440\,in-lb/lf$	[-163 *N-m/m*]
Earth Load	*Unchanged from Construction Case A1; refer to Section* 3.2.1.	

<u>*Parallel Shear*</u>

Unchanged from Construction Case A1; refer to Section 3.2.1.

<u>*Perpendicular Shear*</u>

Unchanged from Construction Case A1; refer to Section 3.2.1.

NOMINAL STRUCTURAL LOAD SUMMARY											
Story	Vertical Location within Wall	Dead		Live		Wind			Earth		
		Axial (plf)	Moment (in-lb/lf)	Axial (plf)	Moment (in-lb/lf)	Moment (in-lb/lf)	Shear,perp (plf)	Shear,par (lb)	Moment (in-lb/lf)	Shear,perp (plf)	Shear,par (lb)
Foundation	Top	800	40	160	-440				0	120	
	Midheight	1,109	20	160	-220				6,278	86	
	x = 3.17 ft	1,223	13	160	-139				7,035	0	
	Bottom	1,419	0	160	0				0	420	N/A

For SI: 1 plf = 14.59390 N/m; 1 in-lb/lf = 0.3706850 N-m/m; 1 lb = 4.448222 N

Figure B3.18 Nominal Load Summary for Construction Case A3

Figure B3.18 is a summary sheet of the loading conditions described above. To determine the moments at various locations in each wall story, refer to Section 7.0. Figure B3.19 is a summary sheet of the ACI [B3] factored loads for each ICF wall story. The values listed in Figure B3.19 are determined by substituting the values from Figure B3.18 into the equations listed in the left column of Figure B3.19.

FACTORED STRUCTURAL LOAD SUMMARY												
FOUNDATION WALL												
ACI 318 Load Combinations[1]	Vertical Location within Wall	Factored Loads										
		Dead		Live		Earth			Total			
		Axial (plf)	Moment (in-lb/lf)	Axial (plf)	Moment (in-lb/lf)	Moment (in-lb/lf)	Shear,perp (plf)	Shear,par (lb)	Axial (plf)	Moment (in-lb/lf)	Shear,perp (plf)	Shear,par (lb)
1. U = 1.4D + 1.7L	Top	1,120	56	272	-748				1,392	-692		
	Midheight	1,553	28	272	-374				1,825	-346		
	x = 3.17 ft	1,712	18	272	-237				1,984	-219		
	Bottom	1,986	0	272	0				2,258	0		
2. U = 1.4D + 1.7L + 1.7H	Top	1,120	56	272	-748	0	204	N/A	1,392	-692	204	N/A
	Midheight	1,553	28	272	-374	10,673	146		1,825	10,327	146	
	x = 3.17 ft	1,712	18	272	-237	11,960	0		1,984	11,740	0	
	Bottom	1,986	0	272	0	0	714		2,258	0	714	
3. U = 0.9D + 1.7H	Top	720	36			0	204	N/A	720	36	204	N/A
	Midheight	998	18			10,673	146		998	10,691	146	
	x = 3.17 ft	1,101	11			11,960	0		1,101	11,971	0	
	Bottom	1,277	0			0	714		1,277	0	714	

[1] D = Dead Load, L = Live Load, W = Wind Load, H = Earth Load

For SI: 1 plf = 14.59390 N/m; 1 in-lb/lf = 0.3706850 N-m/m; 1 lb = 4.448222 N

Figure B3.19 Factored Load Summary for Construction Case A3

Assume the section does not require reinforcement.

Check Perpendicular Shear

According to Figure B3.19, the critical factored perpendicular shear load, V_u, is the same as that in Construction Case A1; refer to Section 3.2.1 for calculations for the structural plain concrete section.

Check Compression and Tension

According to Figure B3.19, the critical maximum total moment, M_u, experienced by the foundation story wall occurs at x = 3.17 feet (0.97 m) due to ACI [B3] Load Combination (3).

The corresponding total factored axial load, P_u, is also taken at x = 3.17 feet (0.97 m) of the foundation wall based on ACI [B3] Load Combination (3). The minimum factored moment required for design is specified in ACI 22.6.3 [B3].

$$M_u = 11{,}971\, in - lb/lf = 1.0\, ft - kip/lf \longleftarrow GOVERNS \quad [4.4\ kN\text{-}m/m]$$
$$P_u = 1{,}101\, plf = 1.1\, kip/lf \quad [16\ kN/m]$$
$$M_{u,min} = 0.1hP_u = 0.1(5.5\, in)(1{,}101\, plf) = 606\, in - lb/lf = 0.05\, ft - kip/lf \quad [225\ N\text{-}m/m]$$

Plot M_u and P_u on the interaction diagram for a 5.5-inch- (140-mm-) thick flat wall shown in Figure B3.20. The plotted point lies below the lower tension boundary for ACI [B3] Load Combination (3); therefore, a 5.5-inch- (140-mm-) thick flat wall structural plain concrete wall is insufficient for the given loading conditions.

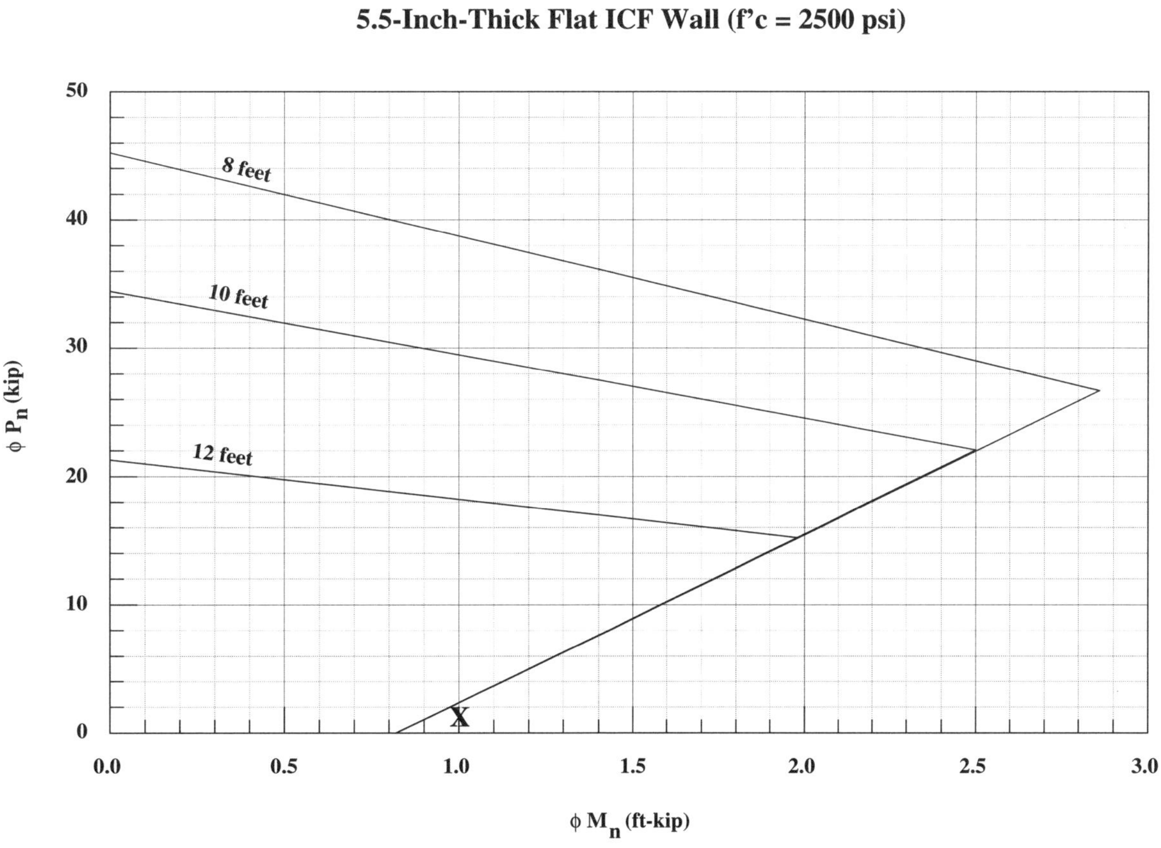

Figure B3.20 Structural Plain 5.5-Inch- (140-mm-) Thick Concrete Wall Interaction Diagram

Design a structurally reinforced wall section with one No. 3 bar at 12 inches (305 mm) on center.

Check Perpendicular Shear

Unchanged from Construction Case A1; refer to Section 3.2.1 for calculations for a structural reinforced section, one No. 3 bar at 12 inches (305 mm) on center.

Determine Sway and Slenderness

Unchanged from Construction Case A1; refer to Section 3.2.1 for calculations for a structural reinforced section, one No. 3 bar at 12 inches (305 mm) on center.

Determine Magnified Moment

With one exception, the equations below are taken from ACI 10.12 [B3]. The equation for *EI*, as listed in ACI 10.12.3 [B3], is applicable to wall sections that contain a double layer of reinforcement. Given that ICFs contain only one layer of reinforcement, the equation for *EI* noted below is used instead [B7].

According to Figure B3.19, the critical maximum total moment, M_u, experienced by the foundation story wall occurs at x = 3.17 feet (0.97 m) due to ACI [B3] Load Combination (3). The corresponding total factored axial load, P_u, is also taken at x = 3.17 feet (0.97 m) of the foundation wall based on ACI [B3] Load Combination (3).

$$M_u = 11{,}971\,in-lb/lf(1\,ft) = 11{,}971\,in-lb \qquad [1.4\ kN\text{-}m]$$

$$P_u = 1{,}101\,plf(1\,ft) = 1{,}101\,lb \qquad [4.9\ kN]$$

$$E_c = 57{,}000\sqrt{f_c'} = 57{,}000\sqrt{2{,}500\,psi} = 2{,}850{,}000\,psi \qquad [19.7\ GPa]$$

$$\beta_d = \frac{P_{u,dead}}{P_u} = \frac{1{,}101\,lb}{1{,}101\,lb} = 1.0$$

$$\rho = \frac{A_s}{A_g} = \frac{0.110\,in^2}{(5.5\,in)(12\,in)} = 0.0017$$

$$\beta = 0.9 + 0.5\beta_d^{\,2} - 12\rho \ge 1.0 = 0.9 + 0.5(1.0)^2 - 12(0.0017) = 1.38$$

$$e = \frac{M_u}{P_u} = \frac{11{,}971\,in-lb}{1{,}101\,lb} = 10.9\ in \qquad [28\ mm]$$

$$EI = \frac{0.4E_cI_g}{\beta} \ge \frac{E_cI_g\left(0.5 - {e}/{h}\right)}{\beta} \ge \frac{0.1E_cI_g}{\beta}$$

$$= \frac{0.1(2{,}850{,}000\,psi)\,{(12\,in)(5.5\,in)^3}/{12}}{1.38} = 34{,}360{,}054\,psi \qquad [237\ GPa]$$

$$M_{2,min} = P_u(0.6 + 0.03h) = 1{,}101\,lb(0.6 + 0.03(5.5\,in)) = 842.3\,in-lb \qquad [95\ N\text{-}m]$$

$$C_m = 1.0\ \textit{for members with transverse loads between supports}$$

$$P_c = \frac{\pi^2\,EI}{(kl_u)^2} = \frac{\pi^2\,(34{,}360{,}054\,psi)}{(1.0)(12\,in/ft)\,(9\,ft)} = 29{,}074\,lb \qquad [129\ kN]$$

$$\delta_{ns} = \frac{C_m}{1-\left(P_u \big/ 0.75P_c\right)} \geq 1.0 = \frac{1.0}{1-\left(1{,}101\,lb \big/ 0.75(29{,}074\,lb)\right)} = 1.053$$

$$M_{ns} = \delta M_u = 1.053(11{,}971\,in-lb) = 12{,}605\,in-lb = 1.05\,ft-kip \qquad [1.4\ kN\text{-}m]$$

Determine Reinforcement

To determine if the wall section is adequately reinforced, plot the magnified moment and the corresponding total factored axial load on an interaction diagram shown in Figure B3.21. The plotted point, "A", lies on the plotted line of the interaction diagram; therefore, one No. 3 bar at 12 inches (305 mm) on center is adequate.

Tests [B6] have shown that horizontal wall reinforcement spacing limited to 48 inches (1.2 m) on center results in reliable performance; therefore, install one No. 4 bar at 48 inches (1.2 m) on center horizontally. At least one continuous horizontal reinforcement bar should be placed within the top 12 inches (305 mm) of the wall story.

Check Deflection

For below-grade walls, a conservative deflection limit of *L/240* for service live loads is used because earth loads are immediate and are not expected to change significantly with time. To calculate wall deflection at service load levels, effective section properties of the assumed cracked concrete section must be established. According to test results [B6], calculating deflection by using $0.1E_cI_g$ was found to be conservative but more accurate than calculating deflection based on the cracking moment per ACI 9.5.2.3 [B3]. When using the moment magnifier concept, it is recommended that the calculated moment magnification factor be applied to the service load moments to be used in conducting the deflection calculations. Assume a gypsum board interior finish exposed to view. For simplicity, when calculating the maximum deflection, assume the earth load acts on the entire wall height.

$$\Delta_{maximum} = \frac{0.01304(\delta)(0.5)ql^5}{0.1E_cI_g} = \frac{0.01304(1.05)(0.5)(30\,pcf)(1\,ft)(9\,ft)^5}{(0.1)(2{,}850{,}000\,psi)\left(\frac{(5.5\,in)(12\,in)^3}{12}\right)}\left(\frac{1{,}728\,in^3}{ft^3}\right) = 0.093\,in \qquad [2.4\ mm]$$

$$\Delta_{allowable} = \frac{L}{240} = \frac{(9\,ft)\left(\frac{12\,in}{ft}\right)}{240} = 0.45\,in \qquad [11.4\ mm]$$

$$\Delta_{actual} \leq \Delta_{allowable} \qquad OK$$

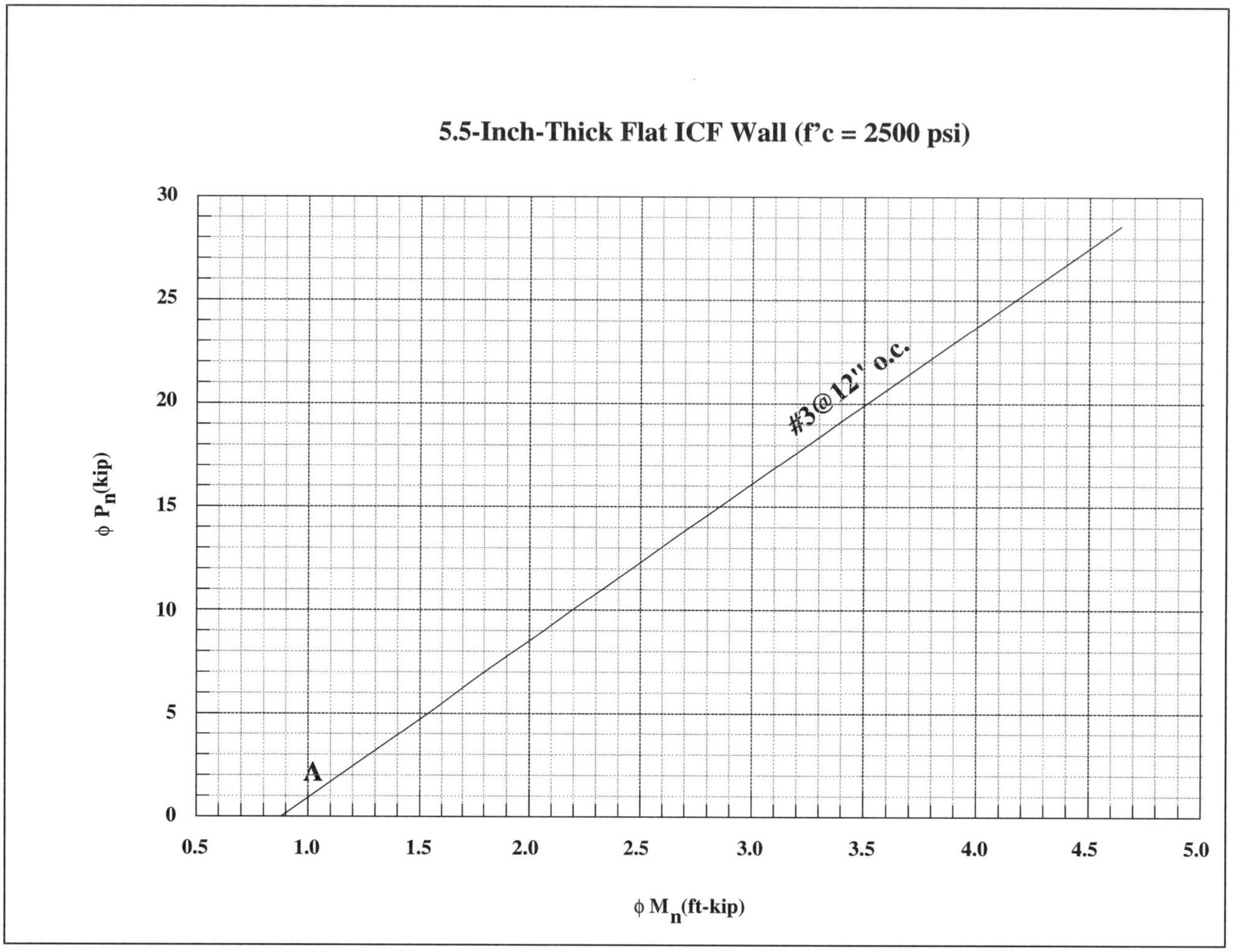

Figure B3.21 Structural Reinforced 5.5-Inch- (140-mm-) Thick Flat Concrete Wall Interaction Diagram

3.2.4 Construction Case A4

Figures B3.22 and B3.23 summarize the nominal and factored loads calculated in this section. Refer to Section 7.0 for typical loading condition equations. The wall moments are calculated by multiplying the axial load by the assumed eccentricities of the roof, wall above, and floor as applicable.

<u>*Dead Loads*</u>

First Floor	$0.5(32\ ft)(15\ psf) = 240\ plf$	[3.5 *kN/m*]
Second Floor	$0.5(32\ ft)(15\ psf) = 240\ plf$	[3.5 *kN/m*]
Roof and Ceiling	$0.5(40\ ft)(16\ psf) = 320\ plf$	[4.7 *kN/m*]
First-Story ICF Wall	$0.5(10\ ft)(100\ psf) = 500\ plf$ *@ mid-height*	[7.3 *kN/m*]
Second-Story ICF Wall	$0.5(10\ ft)(75\ psf) = 375\ plf$ *@ mid-height*	[5.5 *kN/m*]
Foundation ICF Wall	$0.5(9\ ft)(68.75\ psf) = 309\ plf$ *@ mid-height*	[4.5 *kN/m*]

Live Loads

First Floor	$0.5(32\ ft)(40\ psf) = 640\ plf$	[9.3 *kN/m*]
Second Floor	$0.5(32\ ft)(30\ psf) = 480\ plf$	[7.0 *kN/m*]
Roof and Ceiling	$0.5(40\ ft)(49\ psf) = 980\ plf$	[14.3 *kN/m*]
Attic	$0.5(40\ ft)(20\ psf) = 400\ plf$	[5.8 *kN/m*]

Foundation Wall Moments

Dead Load$_{@top}$ $\quad [240\ plf + 2(375\ plf) + 2(500\ plf) + 320\ plf](0\ in)$

$+[240\ plf(-4.75\ in)] = -1{,}140\ in - lb/lf$ [-423 *N-m/m*]

Live Load $_{@top}$ $\quad [(480\ plf + 400\ plf + 980\ plf)](0\ in) + [640\ plf(-4.75\ in)]$

$= -3{,}040\ in - lb/lf$ [1.1 *kN-m/m*]

Earth Load *Unchanged from Construction Case A1; refer to Section* 3.2.1.

Parallel Shear

Unchanged from Construction Case A1; refer to Section 3.2.1.

Perpendicular Shear

Unchanged from Construction Case A1; refer to Section 3.2.1.

NOMINAL STRUCTURAL LOAD SUMMARY

Story	Vertical Location within Wall	Dead		Live		Wind			Earth		
		Axial (plf)	Moment (in-lb/lf)	Axial (plf)	Moment (in-lb/lf)	Moment (in-lb/lf)	$Shear_{,perp}$ (plf)	$Shear_{,par}$ (lb)	Moment (in-lb/lf)	$Shear_{,perp}$ (plf)	$Shear_{,par}$ (lb)
Foundation	Top	2,550	-1,140	2,500	-3,040				0	120	
	Midheight	2,859	-570	2,500	-1,520				6,278	86	
	x = 3.17 ft	2,973	-361	2,500	-962				7,035	0	
	Bottom	3,169	0	2,500	0				0	420	N/A

For SI: 1 plf = 14.59390 N/m; 1 in-lb/lf = 0.3706850 N-m/m; 1 lb = 4.448222 N

Figure B3.22 Nominal Load Summary for Construction Case A4

Figure B3.22 is a summary sheet of the loading conditions described above. To determine the moments at various locations in each wall story, refer to Section 7.0. Figure B3.23 is a summary sheet of the ACI [B3] factored loads for each ICF wall story. The values listed in Figure B3.23 are determined by substituting the values from Figure B3.22 into the equations listed in the left column of Figure B3.23.

Assume the section does not require reinforcement.

Check Perpendicular Shear

According to Figure B3.23, the critical factored perpendicular shear load, V_u, is the same as that in Construction Case A1; refer to Section 3.2.1 for calculations for a structural plain concrete section.

FACTORED STRUCTURAL LOAD SUMMARY														
FOUNDATION WALL														
ACI 318 Load Combinations[1]	Vertical Location within Wall	Factored Loads												
		Dead		Live		Earth			Total			
		Axial (plf)	Moment (in-lb/lf)	Axial (plf)	Moment (in-lb/lf)	Moment (in-lb/lf)	$Shear_{,perp}$ (plf)	$Shear_{,par}$ (lb)	Axial (plf)	Moment (in-lb/lf)	$Shear_{,perp}$ (plf)	$Shear_{,par}$ (lb)
1. U = 1.4D + 1.7L	Top	3,570	-1,596	4,250	-5,168				7,820	-6,764		
	Midheight	4,003	-798	4,250	-2,584				8,253	-3,382		
	x = 3.17 ft	4,162	-505	4,250	-1,636				8,412	-2,141		
	Bottom	4,436	0	4,250	0				8,686	0		
2. U = 1.4D + 1.7L + 1.7H	Top	3,570	-1,596	4,250	-5,168	0	204	N/A	7,820	-6,764	204	N/A
	Midheight	4,003	-798	4,250	-2,584	10,673	146		8,253	7,291	146	
	x = 3.17 ft	4,162	-505	4,250	-1,636	11,960	0		8,412	9,818	0	
	Bottom	4,436	0	4,250	0	0	714		8,686	0	714	
3. U = 0.9D + 1.7H	Top	2,295	-1,026			0	204	N/A	2,295	-1,026	204	N/A
	Midheight	2,573	-513			10,673	146		2,573	10,160	146	
	x = 3.17 ft	2,676	-325			11,960	0		2,676	11,635	0	
	Bottom	2,852	0			0	714		2,852	0	714	

[1] D = Dead Load, L = Live Load, W = Wind Load, H = Earth Load

For SI: 1 plf = 14.59390 N/m; 1 in-lb/lf = 0.3706850 N-m/m; 1 lb = 4.448222 N

Figure B3.23 Factored Load Summary for Construction Case A4

Check Compression and Tension

According to Figure B3.23, the critical maximum total moment, M_u, experienced by the foundation story wall occurs at x = 3.17 feet (0.97 m) due to ACI [B3] Load Combination (3). The corresponding total factored axial load, P_u, is also taken at x = 3.17 feet (0.97 m) of the foundation wall based on ACI [B3] Load Combination (3). The minimum factored moment required for design is specified in ACI 22.6.3 [B3].

$$M_u = 11{,}636\, in-lb/lf = 0.97\, ft-kip/lf \longleftarrow GOVERNS \qquad [4.3\ kN\text{-}m/m]$$

$$P_u = 2{,}676\, plf = 2.7\, kip/lf \qquad [39\ kN/m]$$

$$M_{u,min} = 0.1hP_u = 0.1(5.5\, in)(2{,}676\, plf) = 1{,}472\, in-lb/lf = 0.12\, ft-kip/lf \qquad [0.55\ kN\text{-}m/m]$$

Plot M_u and P_u on the interaction diagram for a 5.5-inch- (140-mm-) thick flat wall shown in Figure B3.24. The plotted point, "X", lies within the lower tension boundary, the upper compression boundary for a 9-foot (2.7-m) wall height, and the reference axes; therefore, a 5.5-inch- (140-mm-) thick flat wall structural plain concrete wall is sufficient for the given loading conditions.

Check Deflection

For below-grade walls, a conservative deflection limit of *L/240* for service live loads is used because earth loads are immediate and are not expected to change significantly with time. To calculate wall deflection at service load levels, effective section properties of the assumed uncracked concrete section are based on E_cI_g. Assume a gypsum board interior finish exposed to view. For simplicity, when calculating the maximum deflection, assume the earth load acts on the entire wall height.

$$\Delta_{maximum} = \frac{0.01304(0.5)ql^5}{E_c I_g} = \frac{0.01304(0.5)(30\ pcf)(1\ ft)(9\ ft)^5}{(2{,}850{,}000\ psi)\left(\frac{(5.5\ in)(12\ in)^3}{12}\right)}\left(\frac{1{,}728\ in^3}{ft^3}\right) = 0.009\ in \qquad [0.2\ mm]$$

$$\Delta_{allowable} = \frac{L}{240} = \frac{(9\ ft)\left(\frac{12\ in}{ft}\right)}{240} = 0.45\ in \qquad [11\ mm]$$

$$\Delta_{actual} \le \Delta_{allowable} \qquad OK$$

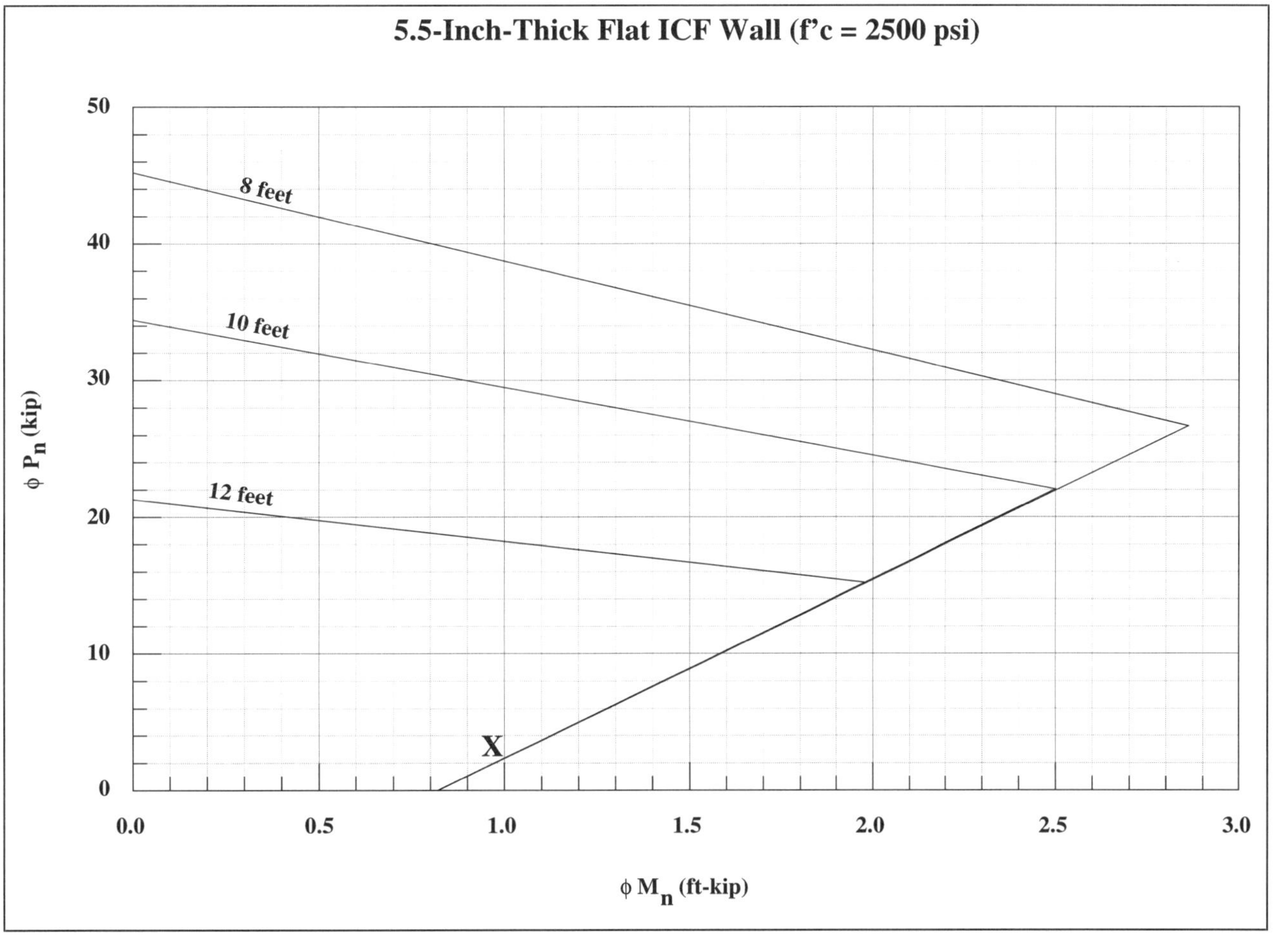

Figure B3.24 Structural Plain 5.5-Inch- (140-mm-) Thick Concrete Wall Interaction Diagram

Determine Reinforcement

Although the wall is designed as a structural plain concrete wall, a nominal amount of reinforcement is typically specified. Tests [B6] have shown that horizontal and vertical wall reinforcement spacing limited to 48 inches (1.2 m) on center results in reliable performance; therefore, install one Grade 40 (300 MPa), minimum No. 4 bar at 48 inches (1.2 m) on center vertically and one No. 4 bar at 48 inches (1.2 m) on center horizontally. At least one continuous horizontal reinforcement bar should be placed within the top 12 inches (305 mm) of the wall story.

3.2.5 Construction Case Summary

Construction Case A1 controls; therefore, a 5.5-inch- (140-mm-) thick flat ICF basement wall that is 9 feet (2.7 m) high with 5 feet (1.5 m) unbalanced backfill with an equivalent fluid density of 30 pcf (481 kg/m^3) requires one No. 3 bar at 10 inches (254 mm) on center minimum vertical wall reinforcement and one No. 4 bar at 48 inches (1.2 m) on center minimum horizontal wall reinforcement. From Table 3.3 of the *Prescriptive Method,* we also obtain one No. 4 bar near third points in the wall story for minimum horizontal wall reinforcement and, from Table 3.4 of the *Prescriptive Method*, one No. 3 bar at 10 inches (254 mm) on center minimum vertical wall reinforcement. It may be more practical to select a larger reinforcement bar size to allow greater spacings and labor savings. For this reason, several design options are provided in Table 3.4.

TABLE 3.3
MINIMUM HORIZONTAL WALL REINFORCEMENT FOR ICF BASEMENT WALLS
(excerpt from the *Prescriptive Method*)

Maximum Height of Basement Wall feet (meters)	Location of Horizontal Reinforcement
8 (2.4)	One No. 4 bar within 12 inches (305 mm) of the top of the wall story and one No. 4 bar near mid-height of the wall story
9 (2.7)	**One No. 4 bar within 12 inches (305 mm) of the top of the wall story and one No. 4 bar near third points in the wall story**

TABLE 3.3
MINIMUM VERTICAL WALL REINFORCEMENT FOR 5.5-INCH- (140-MM-) THICK FLAT ICF BASEMENT WALLS [1,2,3]
(excerpt from the *Prescriptive Method*)

Max. Height of Basement Wall (feet)	Maximum Unbalanced Backfill Height[4] (feet)	Minimum Vertical Reinforcement		
		Maximum Equivalent Fluid Density 30 pcf	Maximum Equivalent Fluid Density 45 pcf	Maximum Equivalent Fluid Density 60 pcf
8	4	#4@48"	#4@48"	#4@48"
	5	#4@48"	#3@12"; #4@22"; #5@32"; #6@40"	#3@8"; #4@14"; #5@20"; #6@26"
	6	#3@12"; #4@22"; #5@30"; #6@40"	#3@8"; #4@14"; #5@20"; #6@24"	#3@6"; #4@10"; #5@14"; #6@20"
	7	#3@8"; #4@14"; #5@22"; #6@26"	#3@5"; #4@10"; #5@14"; #6@18"	#3@4"; #4@6"; #5@10"; #6@14"
9	4	#4@48"	#4@48"	#4@48"
	5	#4@48"	#3@12"; #4@20"; #5@28"; #6@36"	#3@8"; #4@14"; #5@20"; #6@22"
	6	**#3@10"; #4@20"; #5@28"; #6@34"**	#3@6"; #4@12"; #5@18"; #6@20"	#4@8"; #5@14"; #6@16"
	7	#3@8"; #4@14"; #5@20"; #6@22"	#4@8"; #5@12"; #6@16"	#4@6"; #5@10"; #6@12"
	8	#3@6"; #4@10"; #5@14"; #6@16"	#4@6"; #5@10"; #6@12"	#4@4"; #5@6"; #6@8"
10	4	#4@48"	#4@48"	#4@48"
	5	#4@48"	#3@10"; #4@18"; #5@26"; #6@30"	#3@6"; #4@14"; #5@18"; #6@20"
	6	#3@10"; #4@18"; #5@24"; #6@30"	#3@6"; #4@12"; #5@16"; #6@18"	#3@4"; #4@8"; #5@12"; #6@14"
	7	#3@6"; #4@12"; #5@16"; #6@18"	#3@4"; #4@8"; #5@12	#4@6"; #5@8"; #6@10"
	8	#3@4"; #4@8"; #5@12"; #6@14"	#4@6"; #5@8"; #6@12"	#4@4"; #5@6"; #6@8"
	9	#3@4"; #4@6"; #5@10"; #6@12"	#4@4"; #5@6"; #6@8"	#5@4"; #6@6"

For SI: 1 foot = 0.3048 m; 1 inch = 25.4 mm; 1 pcf = 16.0179 kg/m^3

1 Deflection criterion is *L/240,* where *L* is the height of the basement wall in inches.
2 Interpolation shall not be permitted.
3 Walls shall be laterally supported at the top before backfilling.
4 Refer to Section 1.0 for the definition of unbalanced backfill height.

4.0 ICF ABOVE-GRADE WALL DESIGN EXAMPLES AND ENGINEERING CALCULATIONS

The following engineering calculations are based on the application of several recognized engineering standards and specifications. The engineering methods were reviewed by a steering committee of industry experts and professional engineers under sponsorship of the U.S. Department of Housing and Urban Development (HUD), the Portland Cement Association (PCA), and the National Association of Home Builders (NAHB).

4.1 6-Inch- (152-mm-) Thick Waffle-Grid ICF Above-Grade Wall

A waffle-grid ICF above-grade wall is selected from Table 4.3 of the *Prescriptive Method* for above-grade walls constructed in an area where wind speeds are 90 mph (145 km/hr). The above-grade wall is 9 feet (2.7 m) high and supports a light-frame roof only. Table 4.3 shows that the above-grade wall requires one No. 4 bar at 8 feet (2.4 m) on center for vertical wall reinforcement. Calculate the capacity and check the adequacy of the 6-inch- (152-mm-) thick waffle-grid ICF above-grade wall.

Using the values in Figure B1.3 and the material properties in Section 2.0, compute the amount of vertical wall reinforcement required in Construction Cases B1 and B2 for a wall supporting a light-frame roof only.

4.1.1 Construction Case B1

Figures B4.1 and B4.2 summarize the nominal and factored loads calculated in this section. Refer to Section 7.0 for typical loading condition equations. The wall moments are calculated by multiplying the axial load by the assumed eccentricities of the roof, wall above, and floor as applicable.

Dead Loads

Roof and Ceiling $0.5(40\,ft)(12\,psf) = 240\,plf$ [3.5 *kN/m*]

First-Story ICF Wall $0.5(9.0\,ft)(55\,psf) = 248\,plf$ @ *mid-height* [3.6 *kN/m*]

Live Loads

Roof and Ceiling $0.5(40\,ft)(0\,psf) = 0\,plf$ [0 *kN/m*]

Attic $0.5(40\,ft)(0\,psf) = 0\,plf$ [0 *kN/m*]

First-Story Wall Moments

Dead Load $_{@top}$ $[240\,plf + 0\,plf](3\,in) = 720\,in-lb/lf$ [267 *N-m/m*]

Live Load $_{@top}$ $(0\,plf + 0\,plf)(3\,in) = 0\,in-lb/lf$ [0 *N-m/m*]

Wind Load $_{@midht}$ $$\frac{27\,psf\,(9\,ft)^2\,(1\,ft)}{8} = 273\,ft-lb/lf = 3{,}281\,in-lb/lf$$ [1.2 *kN-m/m*]

Parallel Shear

Refer to Section 5.0 for parallel shear calculations.

Perpendicular Shear

Refer to Section 7.0 for variable definitions.

$$V_u = \frac{ql}{2} = \frac{(27\ psf)(9\ ft)}{2} = 121.5\ plf\left(\frac{ft}{1\,vertical\,core}\right) = 122\ lb\,/\,post \qquad [543\ kN/post]$$

NOMINAL STRUCTURAL LOAD SUMMARY											
Story	Vertical Location within Wall	Dead		Live		Wind[1]			Earth		
		Axial (plf)	Moment (in-lb/lf)	Axial (plf)	Moment (in-lb/lf)	Moment (in-lb/lf)	$Shear_{,perp}$ (plf)	$Shear_{,par}$ (lb)	Moment (in-lb/lf)	$Shear_{,perp}$ (plf)	$Shear_{,par}$ (lb)
First	Top	240	720	0	0	0	122	N/A			
	Midheight	488	360	0	0	3281	0				
	Bottom	735	0	0	0	0	122				

[1] Values listed for wind are magnitudes only and therefore are nondirectional.

For SI: 1 plf = 14.59390 N/m; 1 in-lb/lf = 0.3706850 N-m/m; 1 lb = 4.448222 N

Figure B4.1 Nominal Load Summary for Construction Case B1

Figure B4.1 is a summary sheet of the loading conditions described above. To determine the moments at various locations in each wall story, refer to Section 7.0. Figure B4.2 is a summary sheet of the ACI [B3] factored loads for each ICF wall story. The values listed in Figure B4.2 are determined by substituting the values from Figure B4.1 into the equations listed in the left column of Figure B4.2.

FACTORED STRUCTURAL LOAD SUMMARY												
FIRST-STORY WALL												
ACI 318 Load Cases[1]	Vertical Location within Wall	Factored Loads										
		Dead		Live		Wind[2]			Total[3]			
		Axial (plf)	Moment (in-lb/lf)	Axial (plf)	Moment (in-lb/lf)	Moment (in-lb/lf)	$Shear_{,perp}$ (plf)	$Shear_{,par}$ (lb)	Axial (plf)	Moment (in-lb/lf)	$Shear_{,perp}$ (plf)	$Shear_{,par}$ (lb)
1. U = 1.4D + 1.7L	Top	336	1,008	0	0				336	1,008		
	Midheight	683	504	0	0				683	504		
	Bottom	1,029	0	0	0				1,029	0		
2. U = 0.75(1.4D + 1.7L +/- 1.7W)	Top	252	756	0	0	0	155	N/A	252	756	155	N/A
	Midheight	512	378	0	0	4,183	0		512	4,561	0	
	Bottom	772	0	0	0	0	155		772	0	155	
3. U = 0.9D +/- 1.3W	Top	216	648			0	158	N/A	216	648	158	N/A
	Midheight	439	324			4,265	0		439	4,589	0	
	Bottom	662	0			0	158		662	0	158	

[1] D = Dead Load, L = Live Load, W = Wind Load, H = Earth Load
ACI 318 Load Case equations for first- and second-story walls are modified to take into account wind forces creating internal positive or negative pressure.
[2] Values listed for wind load are critical magnitudes only and therefore are nondirectional.
[3] Values listed for total load are critical magnitudes only and therefore are nondirectional.

For SI: 1 plf = 14.59390 N/m; 1 in-lb/lf = 0.3706850 N-m/m; 1 lb = 4.448222 N

Figure B4.2 Factored Load Summary for Construction Case B1

Assume the section does not require reinforcement.

Check Perpendicular Shear

According to Figure B4.2, the critical factored perpendicular shear load, V_u, experienced by the first-story wall occurs at the bottom of the wall story due to ACI [B3] Load Combination (3).

$$V_u = 158\, plf \left(\frac{ft}{1\, vertical\ core} \right) = 158\, lb / post$$ [703 *N/post*]

$$\phi V_n = 0.65 \left(\frac{4}{3} \right) \sqrt{f'_c}\, bh = 0.65 \left(\frac{4}{3} \right) \sqrt{2{,}500\, psi}\, (6.25\, in)(5\, in) = 1{,}354\, lb / post$$ [6.0 *kN/post*]

$$V_u \leq \phi V_n \quad OK$$

Check Compression and Tension

According to Figure B4.2, the critical maximum total moment, M_u, experienced by the first-story wall occurs at mid-height due to ACI [B3] Load Combination (2). The corresponding total factored axial load, P_u, is also taken at mid-height of the first-story wall based on ACI [B3] Load Combination (2). The minimum factored moment required for design is specified in ACI 22.6.3 [B3].

$$M_u = 4{,}561\, in - lb / post = 0.38\, ft - kip / post \longleftarrow GOVERNS$$ [0.52 *kN-m/post*]

$$P_u = 512\, lb / post = 0.51\, kip / post$$ [2.3 *kN/post*]

$$M_{u,min} = 0.1hP_u = 0.1(5\, in)(512\, plf) = 256\, in - lb / post = 0.03\, ft - kip\ / post$$ [29 *N-m/post*]

Plot M_u and P_u on the interaction diagram for a 6-inch- (152-mm-) thick waffle-grid wall shown in Figure B4.3. The plotted point, "B1", lies within the lower tension boundary, the upper compression boundary for a 9-foot (2.7-m) wall height, and the reference axes; therefore, a 6-inch- (152-mm-) thick waffle-grid structural plain concrete wall is sufficient for the given loading conditions.

Check Deflection

For above-grade walls, a conservative deflection limit of *L/240* for total service loads is used. To calculate wall deflection at service load levels, effective section properties of the assumed uncracked concrete section are based on $E_c I_g$.

Assume a gypsum board interior finish exposed to view. The deflection calculations below are based on wind loads. Refer to Section 7.0 for the maximum deflection equation.

$$\Delta_{actual} = \frac{5(27\, psf)(1\, ft)(9\, ft)^4 \left(\frac{1{,}728\, in^3}{ft^3} \right)}{384(2{,}850{,}000\, psi) \left(\frac{(6.25\, in)(5\, in)^3}{12} \right)} = 0.021\, in$$ [0.53 *mm*]

$$\Delta_{allowable} = \frac{(9\, ft)\left(\frac{12\, in}{ft} \right)}{240} = 0.45\, in$$ [11.4 *mm*]

$$\Delta_{actual} \leq \Delta_{allowable} \quad OK$$

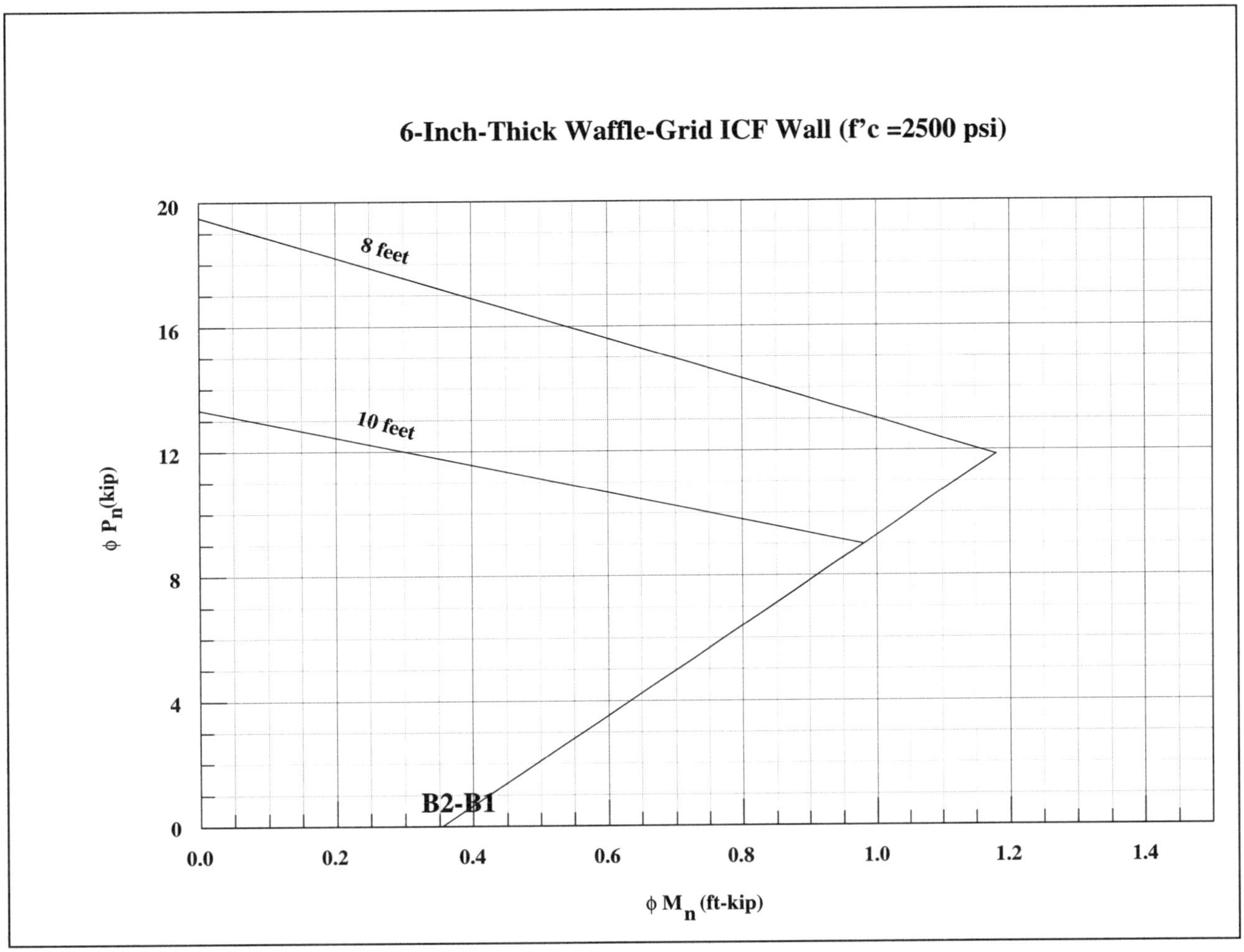

Figure B4.3 Structural Plain 6-Inch- (152-mm-) Thick Concrete Wall Interaction Diagram

<u>Determine Reinforcement</u>

Although the wall is designed as a structural plain concrete wall, a nominal amount of reinforcement is typically specified. Tests [B6] have shown that horizontal wall reinforcement spacing limited to 48 inches (1.2 m) on center results in reliable performance; therefore, install one Grade 40 (300 MPa), minimum one No. 4 bar at 48 inches (1.2 m) on center horizontally. At least one continuous horizontal reinforcement bar should be placed within the top 12 inches (305 mm) of the wall story.

4.1.2 Construction Case B2

Figures B4.4 and B4.5 summarize the nominal and factored loads calculated in this section. Refer to Section 7.0 for typical loading condition equations. The wall moments are calculated by multiplying the axial load by the assumed eccentricities of the roof, wall above, and floor as applicable.

<u>Dead Loads</u>

Roof and Ceiling $0.5(40\,ft)(16\,psf) = 320\,plf$ [4.7 *kN/m*]

First-Story ICF Wall $0.5(9\ ft)(55\ psf) = 248\ plf\ @\ mid\text{-}height$ [3.6 *kN/m*]

<u>*Live Loads*</u>

Roof and Ceiling $0.5(40\ ft)(49\ psf) = 980\ plf$ [14.3 *kN/m*]

Attic $0.5(40\ ft)(20\ psf) = 400\ plf$ [5.8 *kN/m*]

<u>*First-Story Wall Moments*</u>

Dead Load $_{@top}$ $[320\ plf + 0\ plf](0\ in) = 0\ in - lb / lf$ [0 *N-m/m*]

Live Load $_{@top}$ $(980\ plf + 400\ plf)(0\ in) = 0\ in - lb / lf$ [0 *N-m/m*]

Wind Load $_{@midht}$ $\dfrac{27\ psf(9\ ft)^2(1\ ft)}{8} = 273\ ft - lb / lf = 3{,}280\ in - lb / lf$ [1.2 *kN-m/m*]

<u>*Parallel Shear*</u>

Refer to Section 5.0 *for parallel shear calculations.*

<u>*Perpendicular Shear*</u>

Refer to Section 7.0 for variable definitions.

$$V_u = \frac{ql}{2} = \frac{(27\ psf)(9\ ft)}{2} = 121.5\ plf\left(\frac{ft}{1\ vertical\ core}\right) = 122\ lb / post$$ [543 *N/post*]

NOMINAL STRUCTURAL LOAD SUMMARY

Story	Vertical Location within Wall	Dead		Live		Wind[1]			Earth		
		Axial (plf)	Moment (in-lb/lf)	Axial (plf)	Moment (in-lb/lf)	Moment (in-lb/lf)	$Shear_{,perp}$ (plf)	$Shear_{,par}$ (lb)	Moment (in-lb/lf)	$Shear_{,perp}$ (plf)	$Shear_{,par}$ (lb)
First	Top	320	0	0	0	0	122	N/A			
	Midheight	568	0	0	0	3,281	0				
	Bottom	815	0	0	0	0	122				

[1] Values listed for wind are magnitudes only and therefore are nondirectional.

For SI: 1 plf = 14.59390 N/m; 1 in-lb/lf = 0.3706850 N-m/m; 1 lb = 4.448222 N

Figure B4.4 Nominal Load Summary for Construction Case B2

Figure B4.4 is a summary sheet of the loading conditions described above. To determine the moments at various locations in each wall story, refer to Section 7.0. Figure B4.5 is a summary sheet of the ACI [B3] factored loads for each ICF wall story. The values listed in Figure B4.5 are determined by substituting the values from Figure B4.4 into the equations listed in the left column of Figure B4.5.

FACTORED STRUCTURAL LOAD SUMMARY												
FIRST-STORY WALL												
		Factored Loads										
ACI 318 Load Combinations[1]	Vertical Location within Wall	Dead		Live		Wind[2]			Total[3]			
		Axial (plf)	Moment (in-lb/lf)	Axial (plf)	Moment (in-lb/lf)	Moment (in-lb/lf)	$Shear_{perp}$ (plf)	$Shear_{par}$ (lb)	Axial (plf)	Moment (in-lb/lf)	$Shear_{perp}$ (plf)	$Shear_{par}$ (lb)
1. U = 1.4D + 1.7L	Top	448	0	0	0				448	0		
	Midheight	795	0	0	0				795	0		
	Bottom	1,141	0	0	0				1,141	0		
2. U = 0.75(1.4D + 1.7L +/- 1.7W)	Top	336	0	0	0	0	155	N/A	336	0	155	N/A
	Midheight	596	0	0	0	4,183	0		596	4,183	0	
	Bottom	856	0	0	0	0	155		856	0	155	
3. U = 0.9D +/- 1.3W	Top	288	0			0	158	N/A	288	0	158	N/A
	Midheight	511	0			4,265	0		511	4,265	0	
	Bottom	734	0			0	158		734	0	158	

[1] D = Dead Load, L = Live Load, W = Wind Load, H = Earth Load
ACI 318 Load Case equations for first- and second-story walls are modified to take into account wind forces creating internal positive or negative pressure.
[2] Values listed for wind load are critical magnitudes only and therefore are nondirectional.
[3] Values listed for total load are critical magnitudes only and therefore are nondirectional.

For SI: 1 plf = 14.59390 N/m; 1 in-lb/lf = 0.3706850 N-m/m; 1 lb = 4.448222 N

Figure B4.5 Factored Load Summary for Construction Case B2

Assume the section does not require reinforcement.

Check Perpendicular Shear

Unchanged from Construction Case B1, refer to Section 4.1.1 for calculations.

Check Compression and Tension

According to Figure B4.5, the critical maximum total moment, M_u, experienced by the first-story wall occurs at mid-height due to ACI [B3] Load Combination (3). The corresponding total factored axial load, P_u, is also taken at mid-height of the first-story wall based on ACI [B3] Load Combination (3). The minimum factored moment required for design is specified in ACI 22.6.3 [B3].

$$M_u = 4{,}265\,in-lb\,/\,post = 0.36\,ft-kip\,/\,post \longleftarrow GOVERNS \qquad [1.6\ kN\text{-}m/m]$$

$$P_u = 511\,lb\,/\,post = 0.51\,kip\,/\,post \qquad [7.5\ kN/m]$$

$$M_{u,min} = 0.1hP_u = 0.1(5\,in)(511\,plf) = 256\,in-lb\,/\,post = 0.02\,ft-kip\,/\,post \qquad [95\ N\text{-}m/m]$$

Plot M_u and P_u on the interaction diagram for a 6-inch- (152-mm-) thick waffle-grid wall shown in Figure B4.3. The plotted point, "B2", lies within the lower tension boundary, the upper compression boundary for a 9-foot (2.7-m) wall height, and the reference axes; therefore, a 6-inch- (152-mm-) thick waffle-grid structural plain concrete wall is sufficient for the given loading conditions.

Check Deflection

Unchanged from Construction Case B1, refer to Section 4.1.1 for calculations.

Determine Reinforcement

Unchanged from Construction Case B1, refer to Section 4.1.1 for calculations.

4.1.3 Construction Case Summary

In Construction Cases B1 and B2, a 6-inch- (152-mm-) thick waffle-grid ICF first-story wall that is 9 feet (2.7 m) high in a 90 mph (145 km/hr) wind speed area requires no minimum vertical wall reinforcement and a minimum of one Grade 40 (300 MPa), No. 4 bar at 48 inches (1.2 m) on center horizontally. From Table 4.1 of the *Prescriptive Method*, we obtain one horizontal No. 4 bar near third points in the wall story and, from Table 4.3, one vertical No. 4 bar at 96 inches (2.4 m) on center. The vertical wall reinforcement requirement in Table 4.3 is based on current masonry construction practice in high wind areas and past successful experience.

TABLE 4.1
MINIMUM HORIZONTAL WALL REINFORCEMENT FOR ICF ABOVE-GRADE WALLS
(excerpt from the *Prescriptive Method*)

ICF Wall Type and Minimum Wall Thickness inches (mm)[1]	Maximum Height of Wall Story feet (meters)	Location of Horizontal Reinforcement
Flat 3.5 (89)	8 (2.4)	One No. 4 bar within 12 inches (305 mm) of the top of the wall story and one No. 4 bar near third points in the wall story
	9 (2.7)	One No. 4 bar within 12 inches (305 mm) of the top of the wall story and one No. 4 bar near third points in the wall story
	10 (3.0)	One No. 4 bar within 12 inches (305 mm) of the top of the wall story and one No. 4 bar near fourth points in the wall story
Flat 5.5 (140) or Waffle-Grid 6 (152) or Screen-Grid 6 (152)	8 (2.4)	One No. 4 bar within 12 inches (305 mm) of the top of the wall story and one No. 4 bar near mid-height of the wall story
	9 (2.7)	**One No. 4 bar within 12 inches (305 mm) of the top of the wall story and one No. 4 bar near third points in the wall story**
	10 (3.0)	One No. 4 bar within 12 inches (305 mm) of the top of the wall story and one No. 4 bar near third points in the wall story

[1] Actual thickness is shown for flat walls while nominal thickness is given for waffle- and screen-grid walls. Refer to Section 2.0 for actual waffle- and screen-grid thickness and dimensions.

TABLE 4.3
MINIMUM VERTICAL WALL REINFORCEMENT FOR WAFFLE-GRID ICF ABOVE-GRADE WALLS [1,2,3]
(excerpt from the *Prescriptive Method*)

Max. Wind Speed	Maximum Wall Height per Story	Minimum Vertical Reinforcement					
		Supporting Light-Frame Roof Only		Supporting Light-Frame Second Story and Roof		Supporting ICF Second Story and Light-Frame Roof	
		Minimum Wall Thickness [4] (inches)					
(mph)	(feet)	6	8	6	8	6	8
70	8	N/R	N/R	N/R	N/R	N/R	N/R
	9	N/R	N/R	N/R	N/R	N/R	N/R
	10	N/R	N/R	N/R	N/R	N/R	N/R
80	8	N/R	N/R	N/R	N/R	N/R	N/R
	9	N/R	N/R	N/R	N/R	N/R	N/R
	10	N/R	N/R	N/R	N/R	N/R	N/R
90	8	#4@96"	#4@96"	#4@96"	#4@96"	#4@96"	#4@96"
	9	**#4@96"**	#4@96"	#4@96"	#4@96"	#4@96"	#4@96"
	10	#3@12"; #4@24"; #5@36"	#4@96"	#3@12"; #4@24"; #5@24"	#4@96"	#3@12"; #4@24"; #5@24"	#4@96"
100	8	#4@96"	#4@96"	#4@96"	#4@96"	#4@96"	#4@96"
	9	#3@12"; #4@24"; #5@36"	#4@96"	#3@12"; #4@24"; #5@36"	#4@96"	#3@12"; #4@24"; #5@24"	#4@96"
	10	#3@12"; #4@24"; #5@24"	#4@96"	#3@12"; #4@24"; #5@24"	#4@96"	#3@12"; #4@12"	#4@96"
110	8	#3@12"; #4@24"; #5@36"	#4@96"	#3@12"; #4@24"; #5@36"	#4@96"	#3@12"; #4@24"; #5@24"	#4@96"
	9	#3@12"; #4@24"; #5@24"	#4@96"	#3@12"; #4@24"; #5@24"	#4@96"	#3@12"; #4@12"	#4@96"
	10	#3@12"; #4@12"; #5@24"	#4@96"	Design Required	#4@96"	Design Required	#4@96"

For SI: 1 foot = 0.3048 m; 1 inch = 25.4 mm; 1 mph = 1.6093 km/hr

[1] N/R indicates no vertical wall reinforcement is required.
[2] Deflection criterion is *L/240*, where *L* is the height of the wall story in inches.
[3] Interpolation shall not be permitted.
[4] Nominal thickness is given ; refer to Section 2.0 for actual concrete wall thickness.

5.0 ICF WALL OPENING DESIGN EXAMPLES AND ENGINEERING CALCULATIONS

5.1 Minimum Percentage of Solid Wall Length Along Exterior 6-Inch- (152-mm-) Thick Waffle-Grid ICF Above-Grade Wall

A waffle-grid ICF above-grade wall is selected from Table 4.3 of the *Prescriptive Method* for above-grade walls constructed in an area where wind speeds are 90 mph (145 km/hr) and the Seismic Zone is 1. The above-grade wall is 9 feet (2.7 m) high and supports a light-frame roof only. Table 4.3 shows that the above-grade wall does not require vertical wall reinforcement. Assume the building has a 6:12 roof slope. Table 5.3 of the *Prescriptive Method* shows that a base percentage of solid wall length is 25 percent. Calculate the capacity and check the adequacy of the 6-inch- (152-mm-) thick waffle-grid ICF load-bearing above-grade wall for parallel shear.

The tables in Sections 3.0 and 4.0 are based on ICF walls without door or window openings. This simplified approach rarely arises in residential construction since walls generally contain windows and doors. The amount of openings affects the lateral (racking) strength of the building, particularly for wind and seismic loading conditions. The *Prescriptive Method* provides recommendations for the amount and placement location of additional reinforcement required around openings. It also addresses the minimum amount of solid wall required to resist racking loads from wind and moderate seismic forces (i.e., Seismic Zones 0, 1, and 2).

The values for the base percentage of solid wall length along exterior wall lines listed in Table 5.3 of the *Prescriptive Method* were calculated using the main wind force resisting wind loads in accordance with SBC [B2] and seismic loads in accordance with ASCE 7 [B1]. The ICF wall was checked using resistance models for a building 30 feet by 60 feet (9 m by 18 m).

A shear model assumed all solid wall segments 24 inches (610 mm) or greater in length acted as one continuous solid wall length to resist a concentrated lateral force at the top of the wall. The equations below are taken from ACI 22.5.4 [B3] to check parallel wall shear.

The first-story wall design is based on the wall parallel to the roof trusses because that wall experiences the highest loads per lineal foot of solid wall as a result of the design wind loads. The shear model assumes that the critical factored parallel shear load is divided by the total length of the west wall minus any openings in the wall, which is the portion of the wall available to resist shear.

$$V_{parallel} = 3{,}361\,lb \qquad [15.0\ kN]$$

$$V_u = 1.3(3{,}361\,lb) = 4{,}369\,lb \qquad [19.4\ kN]$$

$$\phi V_n = \phi \frac{4}{3}\sqrt{f'_c}\,bh = 0.65\left(\frac{4}{3}\right)\sqrt{2{,}500\,psi}(5\,in)(6.25\,in) = 1{,}354\,lb/post \qquad [6\ kN]$$

$$V_u = \frac{V_u}{solid\ wall\ length}\left(\frac{ft}{1\,vertical\ core}\right)$$

$$V_u \le \phi V_n$$

$$1{,}354\ lb\,/\,post = \frac{4{,}369\ lb}{(32\ ft - x)}\left(\frac{ft}{1\ vertical\ core}\right)$$

$x = 28.8\ ft\ \ maximum\ opening\ wall\ length$ [8.8 m]

$\therefore 32\ ft - 28.8\ ft = 3.2\ ft\ \ solid\ wall\ length\ = 10\ percent$

Because several separate wall segments of various lengths do not have as much capacity as one continuous solid wall segment to resist lateral loads, the results generated using the shear model were deemed unconservative for conditions where large amounts of openings may be possible (i.e., the building acts more like a concrete frame than a bearing wall system).

A flexure model assumed all solid wall segments 24 inches (610 mm) or greater in length acted separately and were equal in length; thereby uniformly distributing the load to each solid wall segment. Each solid wall segment was assumed to act as an unreinforced fixed-end cantilever, 8 feet (2.4 m) in height, to resist a concentrated lateral load at the top of the wall. Wall openings in residential structures are typically neither 8 feet (2.4 m) in height nor do they create solid wall segments completely unsupported at the top; therefore, several flexure models were investigated for wall segments with solid wall lengths ranging from 2 feet (610 mm) to 12 feet (3.7 m). All solid wall length segments occurring in a structure are rarely equal in length; however, the flexure models provided a range of acceptable values that were more conservative than the shear model discussed previously. The calculations below are shown to check the flexural capacity of a 4-foot- (1.2-m-) long, 8-foot- (2.4-m-) high, fixed-end, cantilevered column that can realistically be expected to occur next to a door opening. We assume an unreinforced wall segment because vertical wall reinforcing is only required to be within 12 inches (305 mm) of each opening. Similar calculations were made for other wall segment lengths although not shown in detail here; however, a discussion on the analysis follows.

$P = 3{,}361\ lb$ [15.0 kN]

$P_u = 1.3(3{,}361\ lb) = 4{,}369\ lb$ [19.4 kN]

$M_u = P_u\,(wall\ segment\ height) = (4{,}369\ lb)(8\ ft) = 34{,}952\ ft - lb$ [47.4 GN]

$\phi\, M_n = \phi(5)\sqrt{f'_c}\,S = 0.65(5)\sqrt{2{,}500\ psi}\,(1{,}395\ in^3) = 226{,}688\ lb$ [1.0 MN]

$$Number\ of\ wall\ segments = \frac{226{,}688\ lb}{34{,}952\ lb} = 1.85$$

The solid wall percent values from the 6-, 8-, 9-, 10-, and 12-foot (1.8-, 2.4-, 2.7-, 3-, and 3.7-m) flexure models were used based on the values from the shear model. For example, as the shear model predicted lesser mounts of solid wall in lower load conditions, a progressively shorter segment flexure model was used. This resulted in a more uniform and logical set of bracing provisions that were more conservative and rational than that derived by the shear wall mode alone. However, judgement was required to achieve the final values. The values were rounded to the nearest 5 percent and are listed in Table 5.3 of the *Prescriptive Method* as base percentages of solid wall length. Additional testing and research is needed in this area to determine more accurate solid wall length percentages when concrete walls contain perforations (i.e., wall openings) of various sizes.

While solid wall lengths less than 24 inches (610 mm) have historically been used in ICF residential construction, these segments are designed with knowledge of the specific load path to the wall segment and may require larger amounts of steel reinforcement. The 24-inch- (610-mm-) minimum solid wall length required at maximum intervals of 18 feet (5.5 m) on center and at all building corners is based on suggestions made by the steering committee, indeterminate wall opening placement locations and sizes, and similar requirements for masonry construction. The required percentage of solid wall length allows parallel shear to be resisted yet provides sufficient area for transmission of axial loads to the foundation. It is the intent that future testing will provide a more economical base percentage of solid wall length. The adjustment factors in Table 5.1 of the *Prescriptive Method* and the guidelines in Table 5.2 of the *Prescriptive Method* used to generate the minimum allowable percentage of solid wall length are based on the effects of building geometry on the wind or seismic loads received along a particular axis (i.e., wall line) of a building.

TABLE 5.3
BASE PERCENTAGE OF SOLID WALL LENGTH ALONG EXTERIOR WALL LINES[1,2,3]
(excerpt from the *Prescriptive Method*)

ICF Wall Type and Minimum Wall Thickness (inches)[4]	Max. Roof Slope	Base Solid Wall Length (percent)									
		Wall Supporting Light-Frame Roof					Wall Supporting ICF or Light-Frame Second Story and Light-Frame Roof				
		Maximum Wind Speed (mph)									
		70	80	90	100	110	70	80	90	100	110
Flat 3.5	3:12	15	15	15	15	20	30	35	40	50	55
	6:12	15	15	20	20	25	30	40	50	55	60
	9:12	20	25	30	40	45	45	60	70	85	95
	12:12	25	35	40	50	60	50	65	80	95	100
Flat 5.5	3:12	15	15	15	15	15	20	25	30	40	40
	6:12	15	15	15	15	20	20	30	35	40	45
	9:12	15	15	20	25	30	35	45	50	60	70
	12:12	20	20	25	35	40	35	50	55	70	75
Flat 7.5	3:12	15	15	15	15	15	20	20	25	30	30
	6:12	15	15	15	15	15	20	20	25	30	35
	9:12	15	15	15	20	25	25	30	40	45	50
	12:12	15	20	20	25	30	30	35	40	50	55
Waffle-Grid 6	3:12	15	15	15	15	20	25	30	35	45	50
	6:12	15	15	**20**	20	25	25	35	**45**	50	55
	9:12	20	20	25	35	40	40	55	60	75	85
	12:12	25	30	35	45	50	45	60	70	85	90
Waffle-Grid 8	3:12	15	15	15	15	15	20	25	30	35	35
	6:12	15	15	15	15	20	20	25	30	35	40
	9:12	15	15	20	25	30	30	40	45	55	60
	12:12	20	20	25	30	35	35	40	50	60	65
Screen-Grid 6	3:12	15	15	20	20	25	30	40	45	55	60
	6:12	15	20	25	30	35	30	40	50	60	70
	9:12	25	30	45	50	65	50	65	80	90	100
	12:12	35	40	55	65	80	55	70	85	100	100
Seismic Zone 2											
Flat, 3.5	N/A	20					35				
Flat, 5.5	N/A	15					30				
Flat, 7.5	N/A	15					25				
Waffle-Grid, 6	N/A	20					35				
Waffle-Grid, 8	N/A	20					30				
Screen-Grid, 6	N/A	25					45				

For SI: 1 inch = 25.4 mm; 1 mph = 1.6093 km/hr

1 Linear interpolation between roof slopes shall be permitted.
2 Base percentages are applicable for maximum 10-foot (3.0-m) wall story heights.
3 N/A indicates not applicable.
4 Actual thickness is shown for flat walls while nominal thickness is given for waffle- and screen-grid walls. Refer to Section 2.0 for actual waffle- and screen-grid thickness and dimensions.

5.2 ICF Lintel Design Examples And Engineering Calculations

The following engineering calculations are based on the application of several recognized engineering standards and specifications. The engineering methods were reviewed by a steering committee of industry experts and professional engineers under sponsorship of the U.S. Department of Housing and Urban Development (HUD), the Portland Cement Association (PCA), and the National Association of Home Builders (NAHB).

5.2.1 Waffle-Grid ICF Lintel Design in a Load-Bearing ICF Wall

A waffle-grid ICF lintel is selected from Table 5.8 of the *Prescriptive Method* for lintels supporting light-frame roofs and subjected to a 30 psf (1.4 kPa) ground snow load. The lintel's nominal thickness is 6 inches (152 mm), with a depth of 20 inches (508 mm). Table 5.8 shows that the lintel has a maximum clear span of 11'-0" (3.35 m). Calculate the capacity and check the adequacy of the 6-inch x 20-inch (152-mm x 508-mm) waffle-grid ICF lintel.

5.2.1.1 Maximum Allowable Span Due to Bending Moment

$$M_u = \phi M_n$$

$$\phi M_n = \phi A_s f_y \left(d - \frac{a}{2} \right)$$

$$\phi = 0.9 \ \textit{(strength reduction factor)}$$

Calculate the reinforcement ratio for one No. 5 bottom bar.

$$\rho = \frac{A_s}{A_c} = \frac{0.31\ in^2}{(5\,in)(4\,in) + (2\,in)(13\ in) + (5\ in)(3\,in - 2\,in)} = 0.0061$$

$$\rho_b = \frac{0.85 f'_c \beta_1}{f_y}\left(\frac{87{,}000}{f_y + 87{,}000}\right) = \frac{0.85(2{,}500\,psi)(0.85)}{40{,}000\,psi}\left(\frac{87{,}000}{40{,}000\,psi + 87{,}000}\right) = 0.0309$$

$$\rho_{max} = 0.75\rho_b = 0.75(0.0309) = 0.0232$$

$$\rho_{min} = 0.0012$$

$$\textit{Since}\ \rho_{max} \le \rho_b \le \rho_{min}\ \textit{OK.}$$

Calculate the depth of the compressive stress block.

$$a = \frac{A_s f_y}{0.85 f'_c b} = \frac{(0.31\,in^2)(40\,ksi)}{(0.85)(2.5\,ksi)(5\,in)} = 1.17\,in \qquad [29.7\ mm]$$

Calculate the nominal moment strength.

$$M_n = A_s f_y \left(d - \frac{a}{2} \right) = (0.31\,in^2)(40\ ksi)\left(18\,in - \frac{1.17\,in}{2} \right) = 216\,in - kip \qquad [24.4\ kN\text{-}m/m]$$

Calculate the factored bending moment.

$$M_u = \phi M_n = (0.9)(216\, in-kip)/12 = 16.2\ ft-kip \qquad [22\ kN\text{-}m]$$

Calculate the load on the lintel.

Live Loads			
	Snow = (0.7)(30 *psf*)	= 21 *psf*	[1.0 *kPa*]
	Attic	= 20 *psf*	[0.96 *kPa*]
	Total Live	= 41 *psf*	[2.0 *kPa*]
Dead Loads	*Roof*	= 15 *psf*	[0.72 *kPa*]
	Total Dead	= 15 *psf*	[0.72 *kPa*]

Factored Load = ((41 *psf*)(1.7) + (15 *psf*)(1.4))(32 *ft*)/(1000)(2) + (20 *in*/16 *in*)(0.500 ft^2)(0.150 kcf)(1.4) =1.58 *klf* [23 *kN/m*]

Note: 0.50 sf (0.05 m^2) of concrete fills one linear foot of waffle-grid form.

Calculate the allowable span; assume both ends are fixed.

$$M = \frac{wl^2}{12} \Rightarrow l = \sqrt{\frac{12M}{w}} = \sqrt{\frac{(12)(16.2\ ft-kip)}{1.58\ klf}} = 11\ ft = 11'-0'' \qquad [3.4\ m]$$

5.2.1.2 Maximum Allowable Span Due to Deflection

Calculate the load on the lintel.

Live Loads			
	Snow = (0.7)(30 *psf*)	= 21 *psf*	[1.0 *kPa*]
	Attic	= 20 *psf*	[0.96 *kPa*]
	Total Live	= 41 *psf*	[2.0 *kPa*]
Dead Loads	*Roof*	= 15 *psf*	[0.72 *kPa*]
	Total Dead	= 15 *psf*	[0.72 *kPa*]

Unfactored load (used in deflection calculations)

=(41 *psf* + 15 *psf*)(32 *ft*)/(2) + (20 *in*/16 *in*)(0.50 *sf*)(150 *pcf*) = 990 *plf* [14.4 *kN/m*]

Deflection limit of lintel

$$\Delta = l/240$$

Deflection of lintel

$$\Delta = \frac{5wl^4}{(0.1)384EI}$$

Calculate moment of inertia, *I*.

Calculate $\bar{y}$ from bottom of lintel

$$\bar{y} = \frac{\Sigma A_i \bar{y}_i}{\Sigma A_i} = \frac{(5\,in)(4\,in)(18\,in) + (2\,in)(13\,in)(9.5\,in) + (5\,in)(3\,in)(1.5\,in)}{(5\,in)(4\,in) + (2\,in)(13\,in) + (5\,in)(3\,in)} = 10.32\,in \qquad [262\ mm]$$

Calculate I.

$$I = \Sigma\left(\frac{b_i h_i^3}{12} + Ad^2\right) = \left(\begin{array}{l} \frac{(5\,in)(4\,in)^3}{12} + (5\,in)(4\,in)(7.7\,in)^2 \\ + \frac{(2\,in)(13\,in)^3}{12} + (2\,in)(13\,in)(0.82\,in)^2 \\ + \frac{(5\,in)(3\,in)^3}{12} + (5\,in)(3\,in)(8.82\,in)^2 \end{array}\right) = 2{,}768\,in^4 \qquad [11.5\ dm^4]$$

Calculate the allowable span.

$$\frac{l}{240} = \frac{5wl^4}{(0.1)384EI} \Rightarrow l = \sqrt[3]{\frac{(0.1)384EI}{5(240)w}} =$$

$$= \sqrt[3]{\frac{(0.1)(384)(3{,}122{,}000\ psi)(2{,}768\,in^4)}{(5)(240)(990\ plf)(144\,in^2)}} = 12.5\,ft = 12' - 6'' \qquad [3.8\ m]$$

Since $12' - 6'' > 11' - 0''$, *bending moment governs span.*

5.2.1.3 Stirrup Design

$$V_u = \phi V_n$$
$$V_n = V_c + V_s$$
$$\phi = 0.85 \ \textit{(strength reduction factor)}$$

Determine the maximum shear force.

$$V_u = \frac{wl}{2} = \frac{(1.58\ klf)(11.0\ ft)}{2} = 17.4\ kips \qquad [77.4\ kN]$$

Determine if stirrups are required.

$$V_c = 2\sqrt{f_c'}b_w d = 2\sqrt{2{,}500\ psi}\,(2\,in)(18\,in) = 3.6\ kips \qquad [16\ kN]$$

$$\frac{1}{2}\phi V_C = (0.85)(3.6\ kip)/2 = 1.5\ kips \qquad [6.7\ kN]$$

Since 17.4 kips > 1.53 kips, stirrups are required.

Check required spacing of No. 3 vertical stirrup (A_v = 0.22 in^2 (142 mm^2)).

$$\text{Required } s = \frac{A_v f_y d}{V_u} = \frac{(0.22\,in^2)(40\,ksi)(18\,in)}{(17.4\,kip - 3.6\,kip)} = 11.5\,in \qquad [292\ mm]$$

Maximum allowable spacing is $\frac{d}{2} = \frac{18\,in}{2} = 9\,in$ [229 *mm*]

Since 9 in < 11.5 in., the maximum allowable spacing governs. Recent testing [B8] demonstrates that the shear reinforcement (i.e., stirrups) requirements in ACI [B3] are excessive for this type of application; however, more testing is required to determine what reductions may apply.

5.2.1.4 Governing Design

A 6-inch x 20-inch (152-mm x 508-mm) waffle-grid ICF lintel in a load-bearing wall with one No. 5 bottom bar may span a maximum of 11'-0" (3.4 m) due to bending moment and requires No. 3 stirrups at a maximum spacing of 9 inches (229 mm) on center. From Table 5.8 of the *Prescriptive Method* we also obtain a maximum clear span of 11'-0" (3.4 m) and two No. 3 stirrups in each vertical concrete core.

TABLE 5.8
MAXIMUM ALLOWABLE CLEAR SPANS FOR
WAFFLE-GRID ICF LINTELS IN LOAD-BEARING WALLS[1,2,3]
NO. 5 BOTTOM BAR SIZE
(excerpt from the *Prescriptive Method*)

Minimum Lintel Thickness, T[4] (inches)	Lintel Depth, D (inches)	Maximum Clear Span					
		Supporting Light-Frame Roof Only		Supporting Light-Frame Second Story and Roof		Supporting ICF Second Story and Light-Frame Roof[5]	
		Maximum Ground Snow Load (psf)					
		30	70	30	70	30	70
6	8	5'-4"	4'-8"	4'-5"	4'-1"	4'-5"	3'-10"
	12	8'-0"	6'-9"	6'-3"	5'-6"	6'-3"	5'-1"
	16	9'-9"	8'-0"	7'-5"	6'-6"	7'-5"	6'-1"
	20	**11'-0"**	9'-1"	8'-5"	7'-5"	8'-5"	6'-11"
	24	12'-2"	10'-0"	9'-3"	8'-2"	9'-3"	7'-8"
8	8	6'-0"	5'-2"	4'-9"	4'-3"	4'-9"	3'-11"
	12	8'-3"	6'-9"	6'-3"	5'-6"	6'-3"	5'-2"
	16	9'-9"	8'-0"	7'-5"	6'-6"	7'-5"	6'-1"

5.2.2 Flat ICF Lintel Design in a Non Load-Bearing ICF Wall

A flat ICF lintel in a non load-bearing wall is selected from Table 5.10 of the *Prescriptive Method* for lintels supporting an ICF second story. The lintel thickness is 5.5 inches (140 mm), with a depth of 12 inches (305 mm). Table 5.10 shows the lintel to have a maximum clear span of 8 feet (2.4 m). Calculate the capacity and check the adequacy of the 5.5-inch x 12-inch (140-mm x 305-mm) flat concrete lintel in a non load-bearing ICF wall.

5.2.2.1 Maximum Allowable Span Due to Bending Moment

$$M_u = \phi M_n$$

$$\phi M_n = \phi A_s f_y \left(d - \frac{a}{2} \right)$$

$$\phi = 0.9 \ \textit{(strength reduction factor)}$$

Calculate the reinforcement ratio for one No. 4 bar horizontal tensile steel.

$$\rho = \frac{A_s}{bd} = \frac{0.2\,in^2}{(5.5\,in)(10\,in)} = 0.0036$$

$$\rho_b = \frac{0.85 f'_c \beta_1}{f_y}\left(\frac{87{,}000}{f_y + 87{,}000}\right) = \frac{0.85(2{,}500\,psi)(0.85)}{40{,}000\,psi}\left(\frac{87{,}000}{40{,}000\,psi + 87{,}000}\right) = 0.0309$$

$$\rho_{max} = 0.75\rho_b = 0.75(0.0309) = 0.0232$$

$$\rho_{min} = 0.0012$$

$$Since\ \rho_{max} \le \rho_b \le \rho_{min}\ OK$$

Calculate the depth of the compressive stress block.

$$a = \frac{A_s f_y}{0.85 f'_c b} = \frac{(0.2\,in^2)(40\,ksi)}{(0.85)(2.5\,ksi)(5.5\,in)} = 0.684\,in \qquad [15.9\ mm]$$

Calculate the nominal moment strength.

$$M_n = A_s f_y \left(d - \frac{a}{2} \right) = (0.20\,in^2)(40\,ksi)\left(10\,in - \frac{0.684\,in}{2}\right) = 77.3\,in - kip \qquad [8.8\ kN\text{-}m]$$

Calculate the factored bending moment.

$$M_u = \phi M_n = (0.9)(77.3\,in - kip)/12 = 5.8\,ft - kip \qquad [7.9\ kN\text{-}m]$$

Calculate the load on the lintel.

Dead Loads	*ICF Wall Above*	= 69 *psf*	[3.3 *kPa*]
	Total Dead	= 69 *psf*	[3.3 *kPa*]

Factored Load = 1.4 ((69 *psf*)(8 *ft*) + (5.5 *in*)(12 *in*)(0.150 *kcf*)/(144 in^2)) = 0.87 *klf* [12.7 *kN/m*]

Calculate the allowable span; assume both ends are fixed.

$$M = \frac{wl^2}{12} \Rightarrow l = \sqrt{\frac{12M}{w}} = \sqrt{\frac{(12)(5.8\ ft-kip)}{0.87\ klf}} = 8.9\ ft = 8'-11'' \qquad [2.7\ m]$$

5.2.2.2 Maximum Allowable Span Due to Deflection

Calculate the load on the lintel.

Dead Load	*ICF Wall Above*	= 69 *psf*	[3.3 *kPa*]
	Total Dead	= 69 *psf*	[3.3 *kPa*]

Unfactored Load (used in deflection calculations)

$$=(69\ psf)(8\ ft) + (5.5\ in)(12\ in)(150\ pcf)/144\ in^2 = 621\ plf \qquad [9.1\ kN/m]$$

Deflection limit of lintel

$$\Delta = \frac{l}{240}$$

Deflection of lintel

$$\Delta = \frac{5wl^4}{(0.1)384EI}$$

Calculate the allowable span.

$$\frac{l}{240} = \frac{5wl^4}{(0.1)384EI} \Rightarrow l = \sqrt[3]{\frac{(0.1)384EI}{5(240)w}}$$

$$= \sqrt[3]{\frac{(0.1)(384)(3{,}122{,}000\ psi)\left(\frac{1}{12}(5.5\ in)(12\ in)^3\right)}{(5)(240)(621\ plf)(144\ in^2)}} = 9.6\ ft = 9'-7'' \qquad [2.9\ m]$$

Since $9'-7'' > 8'-11''$*, bending moment governs span.*

5.2.2.3 Stirrup Design

$$V_u = \phi V_n$$
$$V_n = V_c + V_s$$
$$\phi = 0.85 \text{ (strength reduction factor)}$$

Determine the maximum shear force.

$$V_u = \frac{wl}{2} = \frac{(0.87\ klf)(8.9\ ft)}{2} = 3.9\ kips \qquad [17.3\ kN]$$

Determine if stirrups are required.

$$V_c = 2\sqrt{f'_c}b_w d = 2\sqrt{2{,}500\,psi}(5.5\,in)(10\,in) = 5.5\,kips \qquad [24.5\ kN]$$

$$\frac{1}{2}\phi V_C = (0.85)(5.5\,kip)/2 = 2.3\,kips \qquad [10.2\ kN]$$

Since $3.9\,kips > 2.3\,kips$ *, stirrups are required.*

Check required spacing of No. 3 vertical stirrups ($A_v = 0.22\ in^2\ (142\ mm^2)$).

$$Required\ s = \frac{A_v f_y d}{V_u} = \frac{(0.22\,in^2)(40\,ksi)(10\,in)}{(3.9\,kip)} = 23\,in \qquad [584\ mm]$$

Maximum allowable spacing is $\frac{d}{2} = \frac{10\,in}{2} = 5\,in$ [127 *mm*]

Since $5\,in < 23\,in$*, the maximum allowable spacing governs. Recent testing [B8] demonstrates that the shear reinforcement (i.e., stirrups) requirements in ACI [B3] are excessive for this type of application; however, more testing is required to determine what reductions may apply.*

5.2.2.4 Governing Design

A 5.5-inch x 12-inch (140-mm x 305-mm) flat ICF lintel in a non load-bearing wall with one No. 4 bar may span a maximum of 8'-11" (2.7 m) due to bending moment and requires No. 3 stirrups at a maximum spacing of 5 inches (127 mm) on center. From Table 5.10 of the *Prescriptive Method* we obtain a maximum clear span of 8 feet (2.4 m) and No. 3 stirrups at *d/2*. The values in Table 5.10 have been adjusted to simplify the table for all ICF wall types.

TABLE 5.10
MAXIMUM ALLOWABLE CLEAR SPANS FOR
FLAT ICF LINTELS IN NON-LOAD-BEARING WALLS[1,2,3]
NO. 4 BOTTOM BAR SIZE
(excerpt from the *Prescriptive Method*)

Minimum Lintel Depth, D (inches)	Maximum Clear Span	
	Supporting Light-Frame Gable End Wall	Supporting ICF Gable End Wall[4]
8	15'-3"	6'-1"
12	19'-11"	**8'-4"**
16	24'-2"	10'-7"

6.0 LEDGER BOARD CONNECTION DESIGN EXAMPLES AND ENGINEERING CALCULATIONS

The following engineering calculations are based on the application of several recognized engineering standards and specifications. The engineering methods were reviewed by a steering committee of industry experts and professional engineers under sponsorship of the U.S. Department of Housing and Urban Development (HUD), the Portland Cement Association (PCA), and the National Association of Home Builders (NAHB).

6.1 Ledger Board-Waffle ICF Connection Design

A wood 1.5-inch x 7.25-inch (38-mm x 184-mm) ledger board is attached to a 6-inch (152-mm) waffle-grid wall. Assume a 4-inch- (102-mm-) diameter hole is cut into the form around each bolt and that the bolt length extends to the center of the ICF wall thickness. Assume a 5/8-inch (19-mm) bolt diameter of A36 steel is used with a 1 3/8-inch- (35-mm-) diameter washer. The floor joists are 2 feet (0.61 m) on center and have a clear span of 22 feet (6.7 m). Wood member and connection design is in accordance with *NDS* [B4].

6.1.1 Calculate Loads

Nominal Service Load

$V = Dead\ Load + Live\ Load$

$V = (0.5)(22\ ft)((40\ psf + 15\ psf)) = 605\ plf$ [8.8 *kN/m*]

Factored Load

$V_u = 1.4\ Dead\ Load + 1.7\ Live\ Load$

$V_u = (0.5)(22\ ft)(1.4(15\ psf) + 1.7(40\ psf)) = 979\ plf$ [14.3 *kN/m*]

6.1.2 Determine Maximum Bolt Spacing Due to Shear-Friction in Concrete

$$V_n = A_{bolt} F_y \mu \le \begin{Bmatrix} 0.2 f'_c A_{concrete} \\ 800 A_{concrete} \end{Bmatrix}$$

$$V_n = \frac{\pi (1.375\ in)^2}{4}(36{,}000\ psi)(0.6) = 32{,}074\ lb$$ [143 *kN*]

$$V_{n,max} = 0.2(2{,}500\ psi)\left(\frac{\pi(4\ in)^2}{4}\right) = 6{,}283\ lb \quad \longleftarrow \quad GOVERNS$$ [28 *kN*]

$$V_{n,max} = 800\left(\frac{\pi(4\ in)^2}{4}\right) = 10{,}053\ lb$$ [44.7 *kN*]

$$V_u \le \phi V_n \quad OK$$

$$x = \frac{\phi V_n \mu}{V_u}$$

$$x = \frac{(0.85)(6{,}283\,lb)(12\,in/ft)(0.6)}{(979\,plf)} = 39.3\,in$$ [998 *mm*]

6.1.3 Determine Maximum Bolt Spacing Due to Tension in Concrete (Anchorage Capacity)

$$\phi V_c = \phi(4)(A_v)\sqrt{f_c'}$$

$$V_u \leq \phi V_c \quad OK$$

$$x = \frac{\phi V_c}{0.75\left(\frac{V_u}{0.6} + 1.7(wind\ load)(wall\ height)\right)}$$

$$x = \frac{\phi 4\left(\pi(2.5\ in)^2\right)\sqrt{2{,}500\,psi}\,(12\,inch/ft)}{0.75\left(\frac{979\ plf}{0.6} + 1.7(40\,psf)(10\,ft)\right)} = 23.1\,in$$ [587 *mm*]

6.1.4 Determine Maximum Bolt Spacing Due to Tension in Bolt Due to Shear-Friction and Wind Suction Pressure

$$T = 0.75\left(\frac{V_u}{0.6} + 1.7(wind\ load)(wall\ height)\right)(bolt\ spacing)$$

$$f_t = \frac{T}{A_{bolt}}$$

$$f_t \leq F_t$$

$$x = \frac{F_t A_b (12\,inches/ft)}{0.75\left(\frac{V_u}{0.6} + 1.7(wind\ load)(wall\ height)\right)}$$

$$x = \frac{(19{,}100\,psi)(0.306\ in^2)(12\,inches/ft)}{0.75\left(\frac{979\ plf}{0.6} + 1.7(40\,psf)(10\,ft)\right)} = 40.5\ in$$ [1.03 *m*]

6.1.5 Determine Maximum Bolt Spacing Due to Shear in Bolt

$$Z_{actual} \leq Z_{allowable}$$

$$Z_{actual} = V(bolt\ spacing)$$

$$Z_{allowable} = 520\,lb/bolt$$

$$x = \frac{(12\,inches/ft)(520\,lb/bolt)}{605\,plf} = 10.3\,in$$ [262 *mm*]

6.1.6 Determine Maximum Bolt Spacing Due to Bending About Strong Axis in Ledger Board

$$f_b = \frac{M}{S_{xx}}$$

$$M = \frac{PL(bolt\ spacing)}{4}$$

$$f_b \le F_b$$

$$x = \frac{4F_b S_{xx}}{V}$$

$$x = \frac{4(850\ psi)(13.14\ in^3)(12\ in/ft)}{605\ plf\,(2\ ft\ joist\ spacing)} = 443.1\ in \qquad [11.3\ m]$$

6.1.7 Determine Maximum Bolt Spacing Due to Bending About Weak Axis in Ledger Board Due to Wind Suction Pressure

$$f_b = \frac{M}{S_{yy}}$$

$$M = \frac{w(bolt\ spacing)^2}{8}$$

$$f_b \le F_b$$

$$x = \sqrt{\frac{8F_b' S_{yy}}{(wind\ load)(wall\ height)}}$$

$$x = (12\ in/ft)\sqrt{\frac{8(1{,}564\ psi)(2.72\ in^3)}{(40\ psf)(10\ ft)}} = 110.7\ in \qquad [2.8\ m]$$

6.1.8 Determine Maximum Bolt Spacing Due to Allowable Bearing at Washer Due to Weak Axis Bending

$$f_{c\perp} = \frac{T}{A_{washer}}$$

$$M = \frac{w(bolt\ spacing)^2}{8}$$

$$f_{c\perp} \le F_{c\perp}$$

$$x = \frac{F_{c\perp}'(A_{washer})}{(wind\ load)(wall\ height)}$$

$$A_{washer} = \pi\left(\frac{1.375\ in}{2}\right)^2 - \pi\left(\frac{0.625\ in}{2}\right)^2 = 1.18\ in^2 \qquad [761\ mm^2]$$

$$x = \frac{(506\ psi)(1.18\ in^2)(12\ in/ft)}{(40\ psf)(10\ ft)} = 17.9\ in \qquad [455\ mm]$$

6.1.9 Minimum Bolt Spacings and Edge Distance as Defined by *NDS* [B4].

Minimum bolt spacing	$3d_b$	$3(0.625\,in) = 1.9\ inches$	[48 *mm*]
Minimum edge distance	$4d_b$	$4(0.625\,in) = 2.5\ inches$	[64 *mm*]
Minimum distance between bolts in a row	$3d_b$	$3(0.625\,in) = 1.9\ inches$	[48 *mm*]
Minimum end distance	12 *in*		[305 *mm*]
Minimum distance between rows of bolts	$3d_b$	$3(0.625\,in) = 1.9\ inches$	[48 *mm*]

6.2 Governing Design

A wood 1.5-inch x 7.25-inch (38-mm x 183-mm) ledger board attached to a 6-inch (152-mm) waffle-grid wall with 5/8-inch-(19-mm) diameter bolts and 1 3/8-inch (35-mm) washers for floor joists that span 22 feet (6.7 m) require a maximum spacing of 10.3 inches (262 mm) on center as governed by shear in the bolt. From Table 6.1 of the *Prescriptive Method*, we also obtain a maximum bolt spacing of 10 inches (254 mm) on center for a staggered 5/8-inch-(15.9-mm-) diameter bolted connection or 20 inches (508 mm) on center for a double-bolted connection. Bolt patterns are very important to good practice with wood construction.

TABLE 6.1
FLOOR LEDGER-ICF WALL CONNECTION (SIDE-BEARING CONNECTION) REQUIREMENTS[1,2,3]
(excerpt from the *Prescriptive Method*)

Maximum Floor Clear Span[4] (feet)	Maximum Anchor Bolt Spacing[5] (inches)			
	Staggered 1/2-Inch-Diameter Anchor Bolts	Staggered 5/8-Inch-Diameter Anchor Bolts	Two 1/2-Inch-Diameter Anchor Bolts[6]	Two 5/8-Inch-Diameter Anchor Bolts[6]
8	18	20	36	40
10	16	18	32	36
12	14	18	28	36
14	12	16	24	32
16	10	14	20	28
18	9	13	18	26
20	8	11	16	22
22	7	**10**	14	**20**
24	7	9	14	18
26	6	9	12	18
28	6	8	12	16
30	5	8	10	16
32	5	7	10	14

For SI: 1 foot = 0.3048 m; 1 inch = 25.4 mm

7.0 TYPICAL BEAM LOADING CONDITIONS

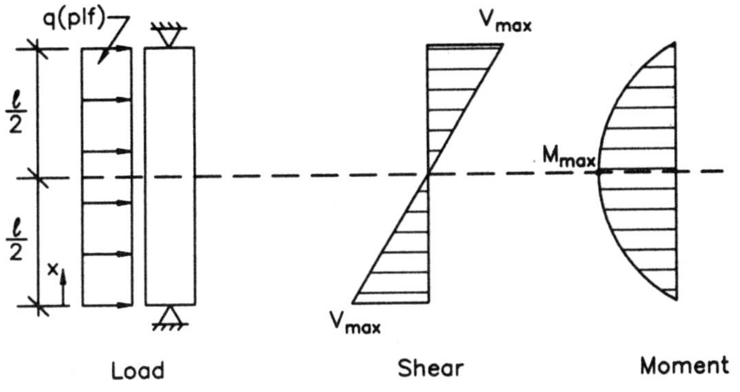

$$V_{max} = \frac{ql}{2}$$

$$V_x = q\left(\frac{l}{2} - x\right)$$

$$M_{max} = \frac{ql^2}{8}$$

$$M_x = \frac{qx}{2}(l - x)$$

$$\Delta_{max@center} = \frac{5ql^4}{384EI}$$

$$\Delta_x = \frac{qx}{24EI}\left(l^3 - 2lx^2 + x^3\right)$$

Figure B7.1 Uniform Load, Simple Span

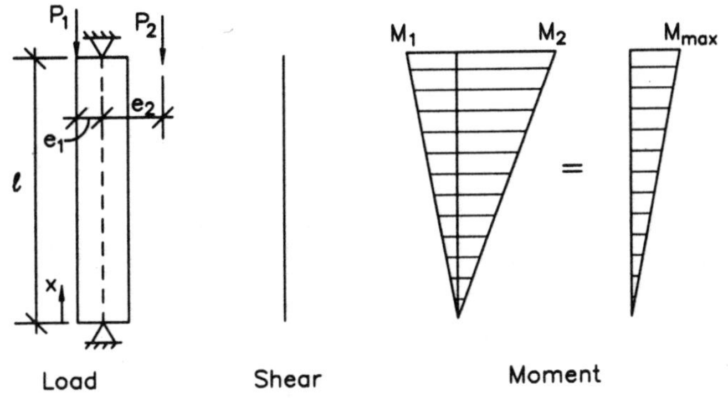

$$V_{max} = V_x = 0$$

$$M_{max} = |M_2| - |M_1| \quad where \; |M_2| > |M_1|$$

$$M_{max} = |M_1| - |M_2| \quad where \; |M_1| > |M_2|$$

$$M_2 = P_2 e_2$$

$$M_1 = P_1 e_1$$

$$M_x = M_{max}\left(\frac{x}{l}\right)$$

Figure B7.2 Eccentric Point Loads, Simple Span

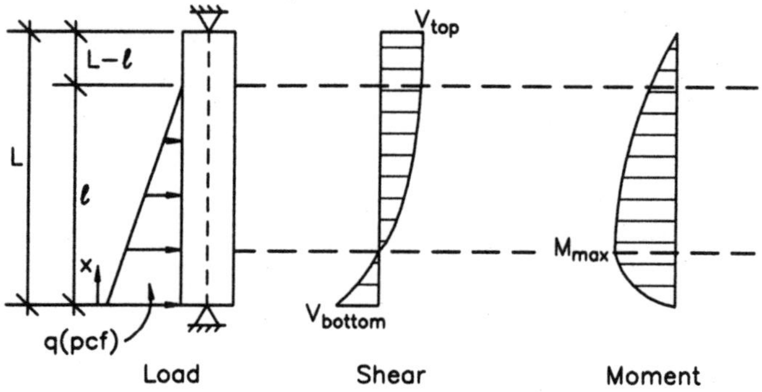

$$V_{top} = \frac{ql^3}{6L}$$

$$V_{bottom} = \frac{ql^2}{2} - \frac{ql^3}{6L}$$

$$V_x = qlx - \frac{qx^2}{2} - V_{bottom} \quad where \quad x < l$$

$$V_x = V_{top} \quad where \quad x \geq l$$

$$M_x = -\frac{qlx^2}{2} + \frac{qx^3}{6} + V_{bottom}x \quad where \quad x < l$$

$$M_x = \frac{ql^3}{6L}(L-x) \quad where \quad x \geq l$$

$$x_{@M_{max}} = \frac{ql - \sqrt{q^2l^2 - 2qV_{bottom}}}{q}$$

Figure B7.3 Partial Triangular Load, Simple Span

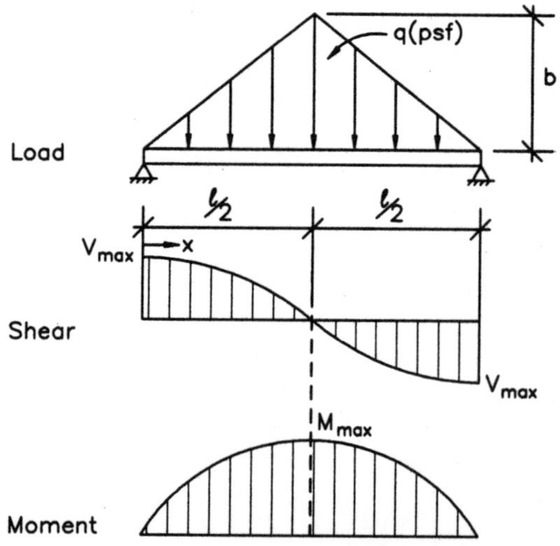

$$V_{max} = \frac{1}{4}bql$$

$$V_x = \frac{bq}{4l}\left(l^2 - 4x^2\right) \quad where \quad x < \frac{l}{2}$$

$$M_{max} = \frac{1}{12}bql^2$$

$$M_x = \frac{bqlx}{2}\left(\frac{1}{2} - \frac{2x^2}{3l^2}\right) \quad where \quad x < \frac{l}{2}$$

$$\Delta_{max\ @\ center} = \frac{bql^4}{120EI}$$

$$\Delta_x = \frac{bqx}{960EI\,l}\left(5l^2 - 4x^2\right)^2 \quad where \quad x < \frac{l}{2}$$

Figure B7.4 Load Uniformly Increasing to Center, Simple Span

8.0 REFERENCES

[B1] *Minimum Design Loads for Buildings and Other Structures* (ASCE 7-93). American Society of Civil Engineers, New York, New York. 1993.

[B2] *Standard Building Code*. Southern Building Code Congress International, Inc. (SBCCI). Birmingham, Alabama. 1994.

[B3] *Building Code Requirements for Structural Concrete* (ACI 318-95). American Concrete Institute, Detroit, Michigan. 1995.

[B4] *National Design Specification for Wood Construction.* The American Forest and Paper Association. Washington, DC. 1997.

[B5] *Manual of Steel Construction Allowable Stress Design, Ninth Edition.* American Institute of Steel Construction. Chicago, Illinois. 1989.

[B6] *Design Criteria for Insulating Concrete Form Wall Systems,* (RP116), Prepared for the Portland Cement Association by Construction Technology Laboratories, Inc., Skokie, Illinois. 1996.

[B7] *Structural Design of Insulating Concrete Form Walls in Residential Construction.* Portland Cement Association, Skokie, Illinois. 1998.

[B8] *Lintel Testing for Reduced Shear Reinforcement in Insulating Concrete Form Systems,* Prepared for the U.S. Department of Housing and Urban Development, Portland Cement Association, and the National Association of Home Builders by the NAHB Research Center, Inc., Upper Marlboro, Maryland. 1998.

APPENDIX C
METRIC CONVERSION FACTORS

The following list provides the conversion relationship between U.S. customary units and the International System (SI) units. A complete guide to the SI system and its use can be found in ASTM E 380, *Metric Practice*.

Length

To convert from	to	Multiply by
inch (in)	meter (m)	2.540 000 E-02
foot (ft)	meter (m)	3.048 000 E-01
yard (yd)	meter (m)	9.144 000 E-01
mile (mi)	meter (m)	1.609 344 E+03

Area

To convert from	to	Multiply by
square foot (ft^2 or sf)	square meter (m^2)	9.290 304 E-02
square inch (in^2)	square meter (m^2)	6.451 600 E-04
square yard (yd^2)	square meter (m^2)	8.361 274 E-01
square mile (mi^2)	square meter (m^2)	2.589 988 E+06

Force per Unit Area (stress or pressure)

To convert from	to	Multiply by
kip per sq. inch (ksi)	Pascal (Pa)	6.894 757 E+06
pound per sq. foot (psf)	Pascal (Pa)	4.788 026 E+01

One Pascal equals 1,000 Newton per square meter.
One kip equals 1,000 pound.

Volume

To convert from	to	Multiply by
cubic inch (in^3)	cubic meter (m^3)	1.638 706 E-05
cubic foot (ft^3)	cubic meter (m^3)	2.831 685 E-02
cubic yard (yd^3)	cubic meter (m^3)	7.645 549 E-01
gallon (gal) Can. liquid	cubic meter (m^3)	4.546 090 E-03
gallon (gal) U.S. liquid	cubic meter (m^3)	3.785 412 E-03
fluid ounce (fl. oz.) U.S. liquid	cubic meter (m^3)	2.957 353 E-05

One U.S. gallon equals 0.8327 Canadian gallon.
One liter equals 0.001 cubic meter.

Force

To convert from	to	Multiply by
kip (1,000 lb)	Newton (N)	4.448 222 E+03
pound (lb)	Newton (N)	4.448 222 E+00
ton (2,000 lb)	Newton (N)	8.896 444 E+03

Force per Unit Length

To convert from	to	Multiply by
kip per linear foot (plf)	Newton per meter (N/m)	1.459 390 E-02
pound per linear foot (plf)	Newton per meter (N/m)	1.459 390 E+01

Mass

To convert from	to	Multiply by
pound (lb), avoirdupois	kilogram (kg)	4.535 924 E-01
ton (2,000 lb)	kilogram (kg)	9.071 847 E+02
slug	kilogram (kg)	1.459 390 E+01

Mass per Unit Length

To convert from	to	Multiply by
kip per linear foot (plf)	kilogram per meter	1.488 164 E-03
pound per linear foot (plf)	kilogram per meter	1.488 164 E+00

Moment

To convert from	to	Multiply by
foot-pound (ft-lb)	Newton-meter (N-m)	1.355 818 E+06

Mass per Unit Volume (Density)

To convert from	to	Multiply by
pound per cubic foot (pcf)	kilogram per cubic meter (kg/m^3)	1.601 846 E+01
pound per cubic yard (lb/yd^3)	kilogram per cubic meter (kg/m^3)	5.932 764 E-01

Velocity

To convert from	to	Multiply by
miles per hour (mph)	kilometer per hour (km/hr)	1.609 344 E+00
miles per hour (mph)	meter per second (m/s)	4.470 400 E-01

Temperature

To convert from	to	Equation
degrees Fahrenheit (°F)	degrees Celsius (°C)	$T_c=(T_f-32)/1.8$
degrees Fahrenheit (°F)	Kelvin (K)	$T_K=(T_f+459.67)/1.8$
Kelvin (K)	degrees Celsius (°C)	$T_c=(T_K-273.15)$

The prefixes and symbols below are commonly used to form names and symbols of the decimal multiples of the SI units.

Multiplication Factor	Prefix	Symbol
$1,000,000,000 = 10^9$	giga	G
$1,000,000 = 10^6$	mega	M
$1,000 = 10^3$	kilo	k

Multiplication Factor	Prefix	Symbol
$0.01 = 10^{-2}$	centi	c
$0.001 = 10^{-3}$	milli	m
$0.000001 = 10^{-6}$	micro	μ
$0.000000001 = 10^{-9}$	nano	n

Reinforcement Bar Data

Inch-Pound	Metric
No. 3	No. 10
No. 4	No. 13
No. 5	No. 16
No. 6	No. 19
No. 7	No. 22
No. 8	No. 25
Grade 40	Grade 300
Grade 60	Grade 420